NEW NARRATIVES IN EIGHTEENTH-CENTURY CHEMISTRY

Archimedes

NEW STUDIES IN THE HISTORY AND PHILOSOPHY OF SCIENCE AND TECHNOLOGY

VOLUME 18

Archimedes has three fundamental goals; to further the integration of the histories of science and technology with one another: to investigate the technical, social and practical histories of specific developments in science and technology; and finally, where possible and desirable, to bring the histories of science and technology into closer contact with the philosophy of science. To these ends, each volume will have its own theme and title and will be planned by one or more members of the Advisory Board in consultation with the editor. Although the volumes have specific themes, the series itself will not be limited to one or even to a few particular areas. Its subjects include any of the sciences, ranging from biology through physics, all aspects of technology, broadly construed, as well as historically-engaged philosophy of science or technology. Taken as a whole, *Archimedes* will be of interest to historians, philosophers, and scientists, as well as to those in business and industry who seek to understand how science and industry have come to be so strongly linked

New Narratives in Eighteenth-Century Chemistry

Contributions from the First Francis Bacon Workshop, 21–23 April 2005, California Institute of Technology, Pasadena, California

Edited by

LAWRENCE M. PRINCIPE

Springer

A C.I.P. Catalogue record for this book is available from the Library of Congress.

ISBN 978-1-4020-6273-5 (HB)
ISBN 978-1-4020-6278-0 (e-book)

Published by Springer,
P.O. Box 17, 3300 AA Dordrecht, The Netherlands.

www.springer.com

Printed on acid-free paper

CONTENTS

THE FRANCIS BACON AWARD

On April 22, 2005 the first Francis Bacon Prize in The History and Philosophy of Science and Technology was awarded to Lawrence M. Principe, Professor of History of Science and Technology and of Chemistry at Johns Hopkins University. Offered biennially by the California Institute of Technology on behalf of the Bacon Foundation in the amount of $20,000, the Prize is awarded to an outstanding scholar whose work has had a substantial impact on the history of science, the history of technology or historically-engaged philosophy of science. As part of the award, Principe was invited to organize a meeting on the history of chemistry with a view to the eventual publication of an edited volume. *New Narratives in Eighteenth Century Chemistry* is the final result of that meeting. It is the first in what we hope will become a series of volumes in *Archimedes* to be edited by the award winners.

Jed Z. Buchwald

NOTES ON CONTRIBUTORS

Bernadette Bensaude-Vincent is professor of the history and philosophy of science at the Université de Paris X. She has authored a number of books about the history and philosophy of chemistry including *A History of Chemistry* (with Isabelle Stengers, 1996) and *Lavoisier: Mémoires d'une revolution* (1993). In 1997, she received the Dexter Award for Outstanding Contributions to the History of Chemistry from the American Chemical Society.

Ku-ming (Kevin) Chang received his Ph.D. at the University of Chicago, and is now assistant professor at the Institute of History and Philology, Academia Sinica, Taiwan. He has published several articles on Georg Ernst Stahl's matter theory, medicine, and alchemy and is revising his dissertation on Stahl's theory of life and matter for publication. He has also published an article that sketches the transformation of the dissertation as a genre of academic publication in early modern Europe, and is embarking on a project that will expand this sketch into a book.

Matthew D. Eddy is lecturer in the history and philosophy of science and an associate of the Centre for the History of Medicine and Disease at the University of Durham. He has most recently held fellowships at the Dibner Institute (MIT), Harvard University, the Max Planck Institute for the History of Science (Berlin), and with the University of Notre Dame's Erasmus Institute. He has written numerous articles on eighteenth- and nineteenth-century intellectual history. Most recently he has edited (with David M. Knight) *Science and Belief: From Natural Philosophy to Natural Science, 1700–1900* (2005) and *William Paley's Natural Theology* (2006). He is currently writing a book on the interactions between medicine, philosophy, and science in Enlightenment Edinburgh.

Ursula Klein is senior research scholar at the Max Planck Institute for the History of Science in Berlin. She is author of *Experiments, Models, Paper Tools: Cultures of Organic Chemistry in the Nineteenth-Century* (Stanford: Stanford University Press, 2003), and (together with Wolfgang Lefèvre) of *Materials in Eighteenth-Century Science: A Historical Ontology* (Cambridge, MA: MIT Press, 2007), as well as editor of *Tools and Modes of Representation in the Laboratory Sciences* (Dordrecht: Kluwer, 2001).

Rina Knoeff is postdoctoral research fellow in the Faculty of Arts at the University of Leiden. She is the author of *Herman Boerhaave (1668–1738). Calvinist Chemist and Physician* (Amsterdam: Edita, 2002). She has worked on the history of medicine and chemistry in relation to (natural) philosophy, theology, and the arts. Currently, she is also involved in research on the Leiden anatomical collections.

Christine Lehman teaches physics and is the author of a Ph.D. dissertation entitled "Gabriel François Venel (1723–1775): Sa place dans la chimie française du XVIIIe siècle" and defended at the Université de Paris X – Nanterre in 2006. Her main research interest is eighteenth-century chemistry.

Trevor H. Levere is University Professor in the Institute for the History and Philosophy of Science in the University of Toronto. His current research is on Dr. Thomas Beddoes (1760–1808) and his circle, and on the interplay between the design of apparatus and the development of concepts in eighteenth- and early nineteenth-century chemistry. He is the author of eight books, including *Discussing Chemistry and Steam: The Minutes of a Coffee House Philosophical Society 1780–1787* (2002), *Transforming Matter: A History of Chemistry from Alchemy to the Buckyball* (2001), *Chemists and Chemistry in Nature and Society 1770–1878* (1994), *Poetry Realized in Nature: Samuel Taylor Coleridge and Early Nineteenth-century Science* (1981), and *Affinity and Matter: Elements of Chemical Philosophy 1800–1865* (1971). Among his five edited books is *Instruments and Experimentation in the History of Chemistry* (2000), with Frederic L. Holmes.

Seymour Mauskopf did his B.A. at Cornell University, and his Ph.D. at Princeton University in the history of science. His fields of research interest are the history of chemistry (*Crystals and Compounds*, 1976; *Chemical Sciences in the Modern World*, 1993) and the history of marginal science (parapsychology) – *The Elusive Science*, with Michael R. McVaugh, 1980. Currently, he is investigating the role of scientists in the development of munitions. In 1998, he received the Dexter Award for Outstanding Contributions to the History of Chemistry from the American Chemical Society. He has taught history of science at Duke University since 1964. He has recently published on Richard Kirwan and the phlogiston theory: "Richard Kirwan's Phlogiston Theory: Its Success and Fate," *Ambix* 49, 2002, 185–205.

John C. Powers is a collateral assistant professor in the Department of History and Assistant Director of the STS Initiative at Virginia Commonwealth University. His current book project, entitled *Inventing Chemistry: Herman Boerhaave and the Reform of the Chemical Arts*, focusses on the shaping of chemistry into a university subject in the early eighteenth century.

Lawrence M. Principe splits his time between the Department of the History of Science and Technology and the Department of Chemistry at Johns Hopkins University, and holds the Drew Professorship of the Humanities. His research interests focus on early modern chymistry, particularly chrysopoeia, and he is currently completing a book provisionally entitled *Wilhelm Homberg and the Transmutations of Chymistry at the Académie Royale des Sciences*. In 2004, he was the inaugural winner of the Francis Bacon Award. His publications include *The Aspiring Adept: Robert Boyle and His Alchemical Quest* (Princeton, 1998) and (with William R. Newman) *Alchemy Tried in the Fire: Starkey, Boyle, and the Fate of Helmontian Chymistry* (Chicago, 2004), the winner of the 2005 Pfizer Prize.

LAWRENCE M. PRINCIPE

A REVOLUTION NOBODY NOTICED?

Changes in Early Eighteenth-Century Chymistry

Among historians of science it is often believed that if there is any century in the history of chemistry about which we are well informed it is the eighteenth. After all, the eighteenth century is home to the Chemical Revolution, an event dating from the last quarter of the century and centered, of course, upon the discoveries and ideas of Antoine Laurent Lavoisier and his colleagues. Lavoisier's work has produced a veritable industry of scholarship; indeed, one that extends well beyond strictly historical studies. There are probably more scholarly words devoted to Lavoisier and developments relating to him than to all the rest of eighteenth-century chemistry put together. In fact, many currently well-known topics in eighteenth-century chemistry have gained attention primarily on account of their eventual relationship to Lavoisier. The clearest example is perhaps the hot topic of combustion, linked especially to Georg Ernst Stahl, his followers, and phlogiston. It has been argued, and convincingly so, that the importance of phlogiston, and even its elaboration as a generalized concept, grew to significant proportions in the eighteenth century only when it became a subject for attack by the "New Chemistry," and consequently that the concept's high profile in historical narratives arrived in part on Lavoisier's coat-tails. It is indisputable that much of the rest of Stahl's chymical system has meanwhile languished in relative neglect. Another example is pneumatic chemistry, linked to Stephen Hales and his successors. This topic too has been given special emphasis because of Lavoisier's eventual success in weighing gaseous products in his analyses and syntheses, and the crucial role that gases played in his work. Thus, much of the history of eighteenth-century chemistry has been written retrospectively, and with the intent of preparing a stage for Lavoisier.[1] Clearly, writing the history of the earlier eighteenth century "backwards" from the later eighteenth century is problematic; it ensures a distorted and highly selective picture. When attempting to focus the messy and often divergent details of historical developments onto a specific target, not only will the relative importance of various topics be determined by their degree of linkage to a particular event and/or character, but entire trains of parallel developments – regardless of their contemporaneous significance – are likely to be neglected.

Lavoisier's is the revolution that everybody has noticed; the one to which I allude in my intentionally provocative title dates from the opposite end of the century. To be sure, I use the term revolution – a term so much abused in the history of

L. M. Principe (ed.), New Narratives in Eighteenth-Century Chemistry: Contributions from the First Francis Bacon Workshop, 21–23 April 2005, 1–22.

science – partly to shock. (I prefer to avoid the generally sterile discussions of what constitutes a revolution and when and how and whether a given one has or has not occurred.) Yet if we simply compare the state of chemistry – its form, aims, practices, and content – in roughly 1675 with its state in about 1725, the differences are so significant and so sudden that they bring the word revolution almost naturally to mind. One might argue in fact that the changes in the discipline over that 50-year period are at least as fundamental as those experienced in an equivalent time–period straddling the celebrated Chemical Revolution – say 1760 to 1810. I am not trying to establish an "alternate revolutionary moment" for the development of chemistry – far from it. Instead, I am merely addressing those who would continue to pitch for the uniqueness (within chemistry) of the rate and profundity of the changes wrought by Lavoisier and his contemporaries and inviting them to consider the changes that took place at other times and over a similar number of years.

Nonetheless, the late seventeenth and early eighteenth century have widely been thought a rather sleepy and uneventful time for chemistry, lying between periods of rapid development occasioned by Boyle at one end and Lavoisier at the other. Although recent work, most particularly that of Frederic L. Holmes and of the scholars whose papers follow in this volume, has begun to alter this perception, as recently as 2002 Robert Siegfried continued to refer to the period from 1675 to 1750 as one of "stagnation."[2] Thus in some quarters at least there remains the image of an eighteenth-century chemistry that is largely "Lavoisier or nothing."

Yet there is much more to the chemistry of the "long" eighteenth century than Lavoisier at one end and Boyle at the other. The need to chart and to better understand this neglected yet crucial period provides the impetus for this volume. In providing an introduction – both for this volume and for the workshop that produced it – I would like to address two main issues. First, I propose that we need to work towards a new, more inclusive, and more accurate organizational narrative (or narratives) for eighteenth-century chemistry. Second, I suggest some features such a narrative should include for the early part of the century, based upon my current research on chymistry at the Académie Royale des Sciences.[3] I do not claim that this list is exhaustive; it represents only some of the important changes of the period that must be incorporated into newer historical accounts.

Undoubtedly some readers will ask if we really need a new organizational narrative. To be sure, such narrative structures – especially when they swell into Grand Narratives with lives of their own – can be more trouble than they are worth. But at the same time, there is room for concern that the very interesting and revealing labors now ongoing might have their impact blunted or diffused if we do not have some larger narrative (though tentative) for the development of chemistry with which to organize them. Part of this concern arises from a more general concern about the current emphasis in the history of science on "local" studies. It is undeniable that careful, detailed, localized studies – what used to be called case studies when they were clearly and convincingly linked to larger questions – are absolutely crucial in order to reveal new features and to avoid incorrect generalizations. But such enthusiasm has to be tempered by the concern that a localized focus that does not speak to larger questions or issues can leave our studies isolated from one another, and unable to contribute meaningfully to a real understanding of historical development. Thus our historical perspective runs the risk of becoming atomized, and thus greatly diminished in explanatory power, and indeed in relevance.

UNSATISFACTORY NARRATIVES FOR EIGHTEENTH-CENTURY CHEMISTRY: THE LAVOISIER-CENTERED AND THE CARTESIAN–NEWTONIAN SHIFT

Before beginning to construct a new narrative, it is helpful to identify the residues and implications of earlier narratives; in other words, to prevent them from misleading or distorting current and future labors. While historiography should never be permitted to substitute for history, a modicum of well-focussed attention to historiography – that is, a careful reflection upon the nature and origins of our own historical assumptions and narratives – can nonetheless be extremely revealing and valuable for better grounding our historical studies.[4] The aforementioned Lavoisier-centered narrative's greatest problem is its teleology, and by consequence, its narrow focus. It thus mediates against the discovery and investigation of issues, characters, locales, and ideas that lie off that well-worn highway running between Boyle and Lavoisier. It makes them seem but a side-show, or even irrelevant. Such a problem is clear enough, and I am not the first to mention it.

There is however a second, equally widespread narrative for eighteenth-century chemistry. This narrative is in fact more germane to my present interests since it deals specifically with the period from around 1675 to 1725. This account claims that the history of chemistry of the period involves the replacement of a largely empirical Aristotelian/Paracelsian seventeenth-century chymistry with a rational Cartesian one that was in turn displaced starting in the early eighteenth century by a Newtonian one. The modern roots of this historiography date largely to the important works of Hélène Metzger in the 1920s and '30s. In her *Doctrines chimiques en France,* Metzger emphasized the role of the pharmacist and lecturer Nicolas Lemery in producing and disseminating a new sort of chemistry with his popular 1675 *Cours de chymie*. Lemery, according to Metzger, was the first to provide clear preparative recipes coupled with mechanical explanations of chemical processes. These mechanical principles Metzger identified as Cartesian in part because they were based upon the shapes, sizes, and motions of minute particles of matter – and they had the important effect of rendering chemistry more rational and orderly than ever before – particularly in contradistinction to a supposedly obscurantist and vitalistic alchemy.[5] In a subsequent work, Metzger then emphasized the role of Newtonianism in bringing about a second fundamental change in chemistry through the concept of forces acting between particles.[6] Thus, Metzger provided a basic narrative for the development of chemistry from the late seventeenth to the mid-eighteenth centuries – first becoming rational by becoming Cartesian, and then developing further towards modernity by becoming Newtonian.

Metzger's narrative was reinforced by Pierre Brunet's nearly contemporaneous presentation of eighteenth-century French science as a battleground between Cartesians and Newtonians.[7] There were of course significant faceoffs between Cartesians and Newtonians, especially in physics, and famously in the issue of the shape of the earth.

But it is much harder to find such a conflict in chemistry. Subsequent developments in the history of science, such as Henry Guerlac's close study of individual Academicians, like Malebranche, have greatly modified the general Cartesian/Newtonian scheme in many contexts, but chemistry is an area where this narrative largely persists to this day.[8] Almost certainly, nationalistic concerns played a role in this perspective, that is, the desire on the part of Anglophones to see the hand of Newton in everything and on the part of Francophones to do the same with Descartes. It was furthered in several writings by Marie Boas and A. Rupert Hall in the 1950s and '60s, and by sections of Arnold Thackray's *Atoms and Powers* in 1970 that asserts the existence within the chemistry of the period of a "Cartesian orthodoxy" and "clandestine Newtonians." Most recently, Mi Gyung Kim's 2003 study, even while it overlays an assemblage of modern jargon and approaches upon the already complex issues of eighteenth-century chemistry, still adopts explicitly and uncritically the old narrative of chemistry passing from empirical Paracelsian to rational Cartesian and thence to Newtonian.[9]

While the Lavoisier-narrative is open to criticism for being narrow, the Cartesian–Newtonian narrative is open to criticism for being Procrustean, if not simply illusory. Early eighteenth-century chemistry fits poorly upon the bed of early eighteenth-century physics (or philosophy), and should not be forced to fit. I argue in fact that we should abandon the generalized notion of Cartesian and Newtonian chemists altogether. Let me make two possibly shocking declarations: First, there were no chemists who were both significant and properly-speaking Cartesian. Second, Isaac Newton is an unnecessary, or at best a marginal term in the mainstream of eighteenth-century chemistry. Let us survey the evidence.

The use of the label Cartesian for chemists and chemistry is now so loose that it means almost nothing – for some French writers it seems to mean no more than "rational" and for English ones it often seems to mean no more than "French." It seems to be applied virtually automatically to almost any system or thinker who employed a concept of shaped corpuscules that give rise mechanically to chemical phenomena. The exemplar of Cartesian chemistry – following the lead of Metzger and reinforced by numerous repetitions since – has been Nicolas Lemery through the vehicle of his enormously popular *Cours de chymie* (first published in 1675).[10] But upon closer examination, it appears that Descartes' thought played little role in Lemery's system.

Lemery's use of mechanical explanations for chemical phenomena is, in fact, very restricted. Although he deploys a particulate matter theory throughout the book, using it both *post* and *ad hoc* to explain chemical processes, he actually talks exclusively of only *two* kinds of shaped particles – acids which are pointy and alkalies which are porous. This system is in marked contrast to that of Descartes who happily postulated shapes for a great variety of particles – smooth, round, straight, bent, branched, hooked, and helical – to explain the properties of everything from oils and salts to magnetic attraction.[11] Moreover, Lemery's explanations based on the shapes of acid and alkali particles alone creates a serious, but hitherto unremarked, tension within his chymical "system." For Lemery equally maintains the traditional five-principle theory (spirit-Mercury, oil-Sulphur, salt, water, and earth) of Étienne de Clave and others, and while his text often uses the relative quantities of these principles to explain the properties of mixed bodies, his mechanical explanations – almost universally sequestered in "Remarques" appended to preparative recipes – *never* engage with the five principles. These two essentially discordant systems exist in parallel spaces, seemingly oblivious of the other's very existence. Thus, the oft-celebrated "mechanical chymical system" of Lemery is nothing of the sort – its scope is restricted, its deployment superficial and uneven.

Lemery's insistence on the shapes of acids and alkalies specifically leads us instead to a different source – namely the chymical acid–alkali theory propounded primarily by Otto Tachenius.[12] In his 1666 *Hippocrates chymicus*, Tachenius claimed that everything is composed of acid and/or alkali, and that the violent reactivity between them explains all chymical and physiological phenomena. It was Tachenius himself who suggested the pointy shape of acid particles and the porous nature of alkali particles used later by Lemery and others. Tachenius based his thinking upon an explicitly sexual metaphor wherein acids are male and alkalies female. Alkalies provide the womb, the vacuity to be filled by phallically-shaped acid particles in a process Tachenius explicitly calls "impregnation." He writes that an alkali being "vacuous and empty," is "impatient of such [emptiness and thus] desires to be saturated with the Acid into Salt, that it may fulfil the course of Nature."[13] He reasons that "if then the *Alcaly* receives, putrefies and cherishes the Acid … it must necessarily perform the Office of a *Mother*, and so be vacuous."[14] This kind of reasoning, though certainly based on the actions of shaped particles, is far indeed from Cartesianism by any standard.

While both Tachenius' role in the acid–alkali theory and the use of this theory by Lemery have previously been recognized,[15] what has *not* been noted is that Tachenius' system, when joined with the corpuscular reasonings present in chymistry from the thirteenth-century Geber down to early seventeenth-century Daniel Sennert and others is sufficient on its own to explain the whole background to Lemery's mechanical explanations, without recourse to Descartes.

As a further non-Cartesian link, just three years before the publication of Lemery's 1675 *Cours*, François de Saint-André, a physician at the Université de Caen, brought out his *Entretiens sur l'acide et l'alkali*, a work that systematized and popularized Tachenius' ideas partly by filtering out the overtly sexual imagery and by applying the theory in a more straightforward and orderly way to chymical phenomena. As a measure of its popularity, Saint-André's book went through three editions in short order.[16] We must then consider a simple question: which is more likely for the apothecary and lecturer Nicolas Lemery to pick up and read in the early 1670s – the 1666 *Hippocrates chymicus* of his celebrated fellow apothecary Otto Tachenius and the 1672 *Entretiens* of Saint-André, a physician and fellow Protestant in Caen, or the philosopher René Descartes' 1647 *Principes de la philosophie*? Although modern historians may have a greater familiarity with the grander figure of Descartes, I think it is safe to argue that a practicing Parisian apothecary of the 1670s was more likely to spend his time amid the pages of Tachenius and Saint-André. It is certainly true that some Cartesian readers (e.g. Bernard de Fontenelle, see below) later found Lemery's explanations harmonious with their own affections for Cartesian principles – and promoted them on that account – but it cannot be said that Cartesianism was Lemery's own source or his inspiration, or that he was engrafting Cartesianism onto the practice of chemistry.

When we move to other important chemists of the period, such as Wilhelm Homberg, the label "Cartesian" which is so regularly, even casually, applied to him makes little sense at all.[17] For it would be a strange sort of Cartesian who believes that light is an emitted stream of material particles, who invokes the *materia subtilis* only rarely in his early career – and even then in modified form – before discarding it, whose substantial published corpus never once mentions a *tourbillon*, who declines the reasoning style of Descartes in favor of sensible chemical demonstrations, and who flatly rejects explicitly Cartesian explanations of phenomena from his contemporaries Johann Bernoulli and Giovanni Domenico Cassini.[18] Indeed, in surveying chemists of this period there is only one who could convincingly be called Cartesian, and that is Louis Lemery, son of the more famous Nicolas, and even here the kind of Cartesianism he adopts is considerably reduced and tailored to more chemical subjects.[19]

As for the term Newtonian, it is as ill-fitting and problematic as Cartesian when applied to eighteenth-century chemistry. Newton's contribution to chemistry has generally been seen in the influence of the "Chymical Queries" that first appeared in the 1706 Latin edition of the *Opticks*. But evidence has been accumulating against a substantial Newtonian contribution to or influence upon early eighteenth-century chemistry. The notion that Étienne-François Geoffroy's 1718 *Table des rapports* embodies Newtonian attractive forces has been well debunked.[20] The famous textbook attributed to Jean-Baptiste Sénac that cites Newton in its title, never applies Newtonian ideas explicitly to any chemical process or observation, and merely provides a summary of the notion of attraction – there referred to as "magnetism" – in an isolated introductory section.[21] Here Newton's name seems to have been more important for advertising than his ideas were for chemical theory, and given what the author actually writes about Newtonian ideas in chemistry, it is clear that he rejects more than he accepts. It was John Freind who applied Newtonian ideas the most clearly and explicitly to chemistry, yet his book had essentially no influence on chemists whatsoever, although it did substantially heat up the Newton–Leibniz feud following Leibniz's negative review of it.[22]

As for Newton's chymical queries themselves, I have elsewhere presented strong evidence that they were directly provoked, if not actually inspired by Wilhelm Homberg's theory of the chymical activity of light, presented publically a year earlier in 1705.[23] Newton's ideas in the first chemical query as published in 1706 bear a striking resemblance to Homberg's. Both systems postulate that light incorporates with matter and serves as its sole source of activity and change. Additionally, Newton's second chymical query – which in the Latin edition was explicitly intended as an illustrative supplement to the first – paraphrases a related publication by Homberg, verifying that Newton was in fact reading Homberg's work at this time. Identifying a source – or at the very least a parallel – for Newton's chymical ideas within French chymistry obviously complicates the issue of isolating a later putative influence of Newton back upon it.

But the whole issue of the introduction of Newtonian forces into chemistry is problematic. Several scholars who have labored on identifying the origins of Newton's ideas of force have pointed to the foregoing chymical tradition.[24] Chymical systems of the seventeenth century – with which Newton was well conversant – frequently endowed matter with active powers, and often explicitly attractive ones, as does for example, Tachenius'. Homberg's system where fixed light is the source of activity in matter is a well-elaborated example of active principles. Yet on the other hand, Newton's ideas of force need not have been borrowed from chymistry at all, given that the attractive and repulsive phenomena of magnetism (for example) provide obvious alternative sources.[25] In either case, the point is that if what are considered to be Newton's characteristic contributions are traceable to earlier traditions, then it becomes considerably more difficult to identify what is and what is not specifically Newton's legacy in later chymical developments; that is, Newton might be considered a superfluous link in the transmission of ideas in chymistry.[26] Thus with few if any truly Cartesian chymists to point to, and little truly Newtonian innovation or influence in chymistry of the first half of the eighteenth century, then the Cartesian–Newtonian narrative cannot be considered historically sound.

CHANGES IN EARLY EIGHTEENTH-CENTURY CHYMISTRY

Now having called former narratives of early eighteenth-century chymistry into question, it is time to say something positive about the period characterized so shockingly in my title as a "revolution nobody noticed." What *are* the significant changes in chymistry during this period? We will need to have these changes – and undoubtedly others as well – in mind when organizing a new narrative for the period. Although a larger number of specific developments might be adduced, I will here cite only the three major features that seem most significant to me at this point. The first involves changes in the aims and applications of chemistry. This feature certainly includes the notable broadening of the usefulness of chemistry for various endeavors but perhaps the most notable single change is the sudden elimination of chrysopoeia (metallic transmutation) from the domain of serious inquiry. A second, and more social yet closely related feature is the increase in the status and professionalization of chemistry and of its practitioners, or at the very least a heightened concern over such status. Third, and somewhat more diffusely, this period saw a remarkable fertility of theoretical innovations based on experimental and practical results. This last feature provides the historian with ample, and in fact still largely untouched material for continuing and extending the reassessment of the complex relations between laboratory practices and theoretical frameworks in chymistry.

A satisfactory exposition of these changes requires more than one book-length treatment, so this opening paper's scope is limited to a single place and a central person that can coherently illustrate all these points at once, thus providing one starting point for further study and discussion. The place is the Parisian Académie

Royale des Sciences, and the person is Wilhelm Homberg (1653–1715). There is little doubt that the Académie was the most important (and best funded) locus for chemistry at the time, and that Homberg was its most important chemist – even if he remains little recognized today – thus this focus is well enough warranted.[27]

The demise of chrysopoeia, and indeed of the search for all the other long-celebrated chymical arcana, is the first dramatic development of this period that I wish to mention. While the pursuit of metallic transmutation and the Philosophers' Stone had been subject to intense debate from the Middle Ages, they remained throughout that time, and down at least to the end of the seventeenth century, plausible or acceptable endeavors among serious natural philosophers and intellectuals. Yet by the 1720s there were no longer any respectible figures visibly involved in chrysopoeia. Kevin Chang, in his contribution to this volume, notes that Stahl's transformation from a believer to a critic of chrysopoeia occurred around 1720, just at the same time chrysopoeia was suddenly vanishing from the French scene. This sudden loss of a pursuit so long-standing and so central to chymistry begs for explanation. The Whiggish notion that as an "error" chrysopoeia was doomed to this fate is unsatisfactory, especially when we can find no refutation of the idea based upon new experiments or new theories.[28] The long-standing reliance within historical accounts upon a dismissive attitude to transmutation has made it seem as if its disappearance was somehow spontaneous – even "overdue." On the contrary, chrysopoeia's disappearance represents a major event in the history of chemistry that needs to be explained. Significantly, transmutational alchemy seems to have retired from the field under an assault of contempt and accusations of fraud rather than under a hail of argument and evidence. And even this demise turns out to be much more complicated than we might expect. I propose that an entrée to an explanation is to be found in the dichotomy between the public and private faces of chemistry at the Académie, and the hitherto largely unrecognized tensions between individual academicians.

I have shown elsewhere that Homberg was deeply involved in studies of metallic transmutation and the search for other chymical arcana throughout his career.[29] In 1684 he worked on a process to transmute mercury into silver. In 1702, when Philippe II, Duc d'Orléans, by then Homberg's tutee in chemistry, purchased the enormous burning lens made by Ehrenfried Walther von Tschirnhaus, the first experiments Homberg performed with it were attempts to use the sun's light to transmute silver into gold. In 1700, Homberg published a paper on a complex purification of mercury which was supposed to give it new properties; I have identified this process as nothing other than the secret preparation of the Philosophical Mercury, the critical starting material for the Philosophers' Stone. Moreover, the specifics of Homberg's process link him inseparably to a main avenue of chrysopoetic research dating back at least to the sixteenth century, and which counted as adherents Alexander von Suchten, Basil Valentine, Johann Joachim Becher, and George Starkey, alias Eirenaeus Philalethes. Indeed, Homberg affirms explicitly that he "followed the entire work of Philalethes" in regard to making Philosophical Mercury and trying to digest it into the Philosophers' Stone.[30] Finally, Homberg's most important publication, his *Essais de chimie*, a kind of a serial textbook published in the Académie's *Mémoires* from 1702 to 1710, is built to a large extent around illustrative experiments with this Philosophical Mercury. Homberg claims that heating in a sealed flask causes the mercury to turn to a black powder which then becomes white and finally red – exactly the sequence of events posited in chrysopoetic texts as guideposts in the successful preparation of the Philosophers' Stone. Homberg also claims that this process converts a portion of the mercury into gold, using a method, which – even though he does not cite it as such in print – was known as the *minera perpetua* by chrysopoeians and described by Johann Joachim Becher, Kenelm Digby, and others. Newly-discovered manuscript material shows unquestionably that Homberg knew and studied chrysopoetic theories extremely well, explicitly considered transmutational alchemy simply as a "part of chemistry," and was willing to believe in the reality of the Philosophers' Stone.[31]

Contemporaries seem to have been aware of Homberg's transmutational interests and endeavors. I recently uncovered a letter that Gottfried Wilhelm Leibniz wrote to Homberg in 1711, asking him to reveal his experiences with the transmutation of metals, noting that such experiments, even if not profitable monetarily, would be useful to refute the erroneous matter theories of Nicolaas Hartsoeker and others.[32] And while it is true that Homberg does not generally reveal his chrysopoetic aspirations explicitly in print, there is clear evidence that some readers recognized them anyway. In a book-length French manuscript dating to about 1720, extant in multiple copies and entitled "Essay to develop the Knowledge and Practice of the Work of the Chemical Philosophers," that is, the Philosophers' Stone, the unidentified author lists several methods for preparing the all-important Philosophical Mercury.[33] After listing the methods of famous authors of seventeenth-century alchemical treatises, such as Pantaleon and Philalethes, he gives also "the method of Mr. Homberg." Even more strikingly, he adopts Homberg's chymical theory, replete with marginal citations to papers in the Académie's *Mémoires*, and grafts this theory seamlessly onto a theory of the Philosophers' Stone and metallic transmutation. This argues still further that early eighteenth-century developments in theory did not defeat chrysopoeia – instead, here a chrysopoeian eagerly adopts the most modern chemical theory in order to *help* in his search. The main point is that Homberg continued to pursue traditional chrysopoeia until his death in 1715 and his colleagues both inside and outside the Académie knew about it.

But Homberg's activities certainly made some in the Académie uneasy. Lemery had lauched a strident attack against gold-making in the third edition of his popular *Cours de chymie*, indicting the entire subject as nothing but fraud, using ridicule as virtually his only weapon.[34] Moreover, while Homberg was pursuing his search for arcana inside the Académie, the institution's public face – largely the construction of the perpetual secretary Bernard de Fontenelle – was heaping scorn such endeavors. Fontenelle already had a low opinion of chymistry in general – seemingly because it could not be reduced to deductive axioms like mathematics and physics – and indeed in his lengthy essay on the utility of the sciences, he mentions *la chimie* only once, and then as no more than an adjunct to medicine.[35] The search for hidden arcana – like transmutation and the alkahest – only made things worse. Indeed, one of Fontenelle's popular *Dialogues of Dead* summons up the shade of Ramon Lull, supposed author of numerous chrysopoetic works, who admits that after his death he finally realized (too late!) that the Philosophers' Stone was a mere chimera.[36] Fontenelle's low opinion of chemistry stands in stark contrast to that of Homberg, who argued that chemistry alone gives sound and sensible accounts about the nature of matter, while the ideas about matter offered by "les physiciens" remain undemonstrable, unprovable, and ultimately detached from sensible reality. Chemistry for Homberg is the most powerful and revealing of the sciences: it has "an infinite extent."[37]

Fontenelle also rarely missed the chance to create a mythological history of chemistry – one that (by no mere coincidence) bears a substantial resemblance to the dismissive historiographies of early chemistry that have only recently been shown incorrect. His main message was that beginning only at the end of the seventeenth century was there any chemical work worthy of the name scientific. He claims that Lemery was "the first who dispersed the natural and affected shadows of chemistry" – despite the fact that Lemery's *Cours de chymie* is really little different either in form, content, or clarity from the already long lineage of chymical – or rather pharmaceutical – recipe books of the French didactic tradition, and as shown above, its most celebrated explanatory principles were borrowed from Tachenius and de Saint-André.[38] Fontenelle clearly found the enormous popularity of Lemery's textbook a useful thing to co-opt. Indeed, Fontenelle's enthusiastic advocacy of Lemery may explain the importance Lemery has gained and maintained to this day; it is likely that Fontenelle's notoriety as a Cartesian popularizer and his early advocacy of Lemery influenced Metzger's own reading of Lemery and his place in the history of chemistry. In fact, however, Lemery's real contributions to chemistry, save as a popularizer, are very slim indeed. Members of the Académie – *pace* Fontenelle – recognized this fact; for instance, as Lemery detailed his exhaustive investigations of the properties of antimony, one academician recorded privately that "all his experiments were found not new in the least, Mr. Homberg has already done most of them."[39] And if there is a significant break in the French textbook tradition, it comes not with Lemery, but only with Homberg and his *Essais de chimie*, which is the first comprehensive text directed specifically towards the development of a foundational, explanatory, and predictive chemical theory based on illustrative experiments rather than simply teaching pharmaceutical recipes.[40]

But why did the Académie, or at least Fontenelle and Lemery, think it so important to denounce transmutation and other arcana, especially when there was no new evidence to back up their claims? Here we approach the second of the important changes I identified in the chemistry of the early eighteenth century – namely the status of chemists and chemistry. In the late seventeenth century, chemistry suffered from a very bad public relations problem. For example, theatregoers to the Paris season of 1693–94 were regaled by *Les Souffleurs, ou la Pierre Philosophale d'Arlequin*, a comedy in the Italian style that depicts a herd of both fools and charlatans, both of them consistently called *chimistes* (the play never uses the word *alchimie* or *alchimiste*). It even contained a satirical song about the marvels and promises of "la chimie."[41] At about the same time, as the records of the Bastille indicate, several chymists were arrested on charges of poisoning, fraud, or counterfeiting. At the start of the period treated here, there was the textbook author and chymical lecturer at the Jardin du Roy, Christophle Glaser, implicated in the celebrated *affaire des poisons*, imprisoned in the Bastille and consequently ruined, but many other unfortunates followed him.[42] In 1712, upon the sudden death of the Dauphin and Dauphine, Homberg himself was implicated as a poisoner and narrowly escaped being taken into the Bastille only by the direct intercession of Louis XIV.[43] The Duc d'Orléans' keen interest in chemistry and his close relationship with Homberg – his chymical tutor, physician, and laboratory collaborator – was well known publically, but this knowledge coexisted with rumors (and perhaps more than just rumors) that the Duc's interest in chymistry was accompanied by one in magic, necromancy, water-gazing, and demonic invocations. At the same time, there was at this time no established place for chymistry at the university which could provide legitimation or visible exemplars.[44] Indeed, the reorganized Académie of 1699 was arguably the first place where permanent positions were earmarked for what we would recognize as chemistry – as distinct from medicine and pharmacy. It is unlikely that the public could distinguish very well between what was being done by the esteemed members of the Académie and what was ridiculed on the stage, declaimed in the popular press, or bruted about in gossip. Thus with the image of chymists as herd of dreamers, fools, mountebanks, and poisoners, serious practitioners in general, and the Académie in particular as an official state institution funded by the Crown, had a serious problem.

The easiest solution, it seems to me – as I believe it seemed to Fontenelle and others – was to make a fresh start for chemistry: to create chemistry afresh as if it had never really existed before. This re-creation of chemistry is an essential part of the "revolution" alluded to in my title – and revolutionaries are notorious for self-serving revisionist histories. By casting the Académie's chemistry as an essentially new subject, and going further by declaiming specifically against the subset of chymistry most easily prey to ill repute – namely chrysopoeia – the Académie could protect the chemistry it was newly professionalizing from the ambiguous status that had dogged *chimia* since the Middle Ages. It was also possible to quarantine all of the questionable activities related to chemistry under a completely different rubric – namely, that of alchemy. Thus it was in this time, place, and period that the previously synonymous words alchemy and chemistry consistently took on their modern meanings for the first time – chemistry as a modern, scientific, and professionalized study, and alchemy as an ancient, unscientific, futile dream and ultimately fraud revolving around imaginary arcana: the Philosophers' Stone, transmutation, the alkahest, and so forth.[45] Of course, chrysopoetic authors had been at pains for centuries to distinguish themselves from frauds, cheats, and other pretenders to the title, but only at the end of the seventeenth and beginning of the eighteenth was there a consistent labeling of *all* chrysopoeia as pure fraud and delusion together with a separation of the chrysopoeians from the rest of their chymical brethren.[46] This rhetoric was particularly strong at scientific societies – particularly the Académie Royale des Sciences, but also to a lesser extent at the Royal Society of London – which were concerned about their public corporate image, and thus the status of the chemistry and chemists in their midst.[47]

This goal lies behind the rhetoric of novelty and the snipes at earlier chymistry that pervade the summaries of academicians' papers that Fontenelle wrote and published as the annual *Histoire* of the Académie. As a point of comparison, the analogous writings of Fontenelle's predecessor as secretary of the Académie, the abbé Jean-Baptiste Duhamel, who apparently had a much higher regard for chymistry, do not share Fontenelle's revisionist and judgmental slant.[48] It also explains the revisionist biographies Fontenelle wrote for Lemery and Homberg. Both chemists supposedly *literally fled* from practitioners of the "old disreputable" chymistry. For Lemery it was from Glaser, whom Fontenelle describes as "a true chemist, full of obscure ideas, greedy of such ideas, and unsociable."[49] For Homberg, his alleged fear over association with a chrysopoetic chymist compelled him leave Paris (with an ingot of gold supposedly made by the unnamed chrysopoeian) and flee to Italy. At this juncture in his Éloge, Fontenelle declares loudly that "Homberg was too capable to aspire to the Philosophers' Stone and too sincere to put such a vain idea into anyone's head."[50] Clearly Fontenelle protests too much, for Homberg himself later freely and unabashedly described in print how at just this time he was busily trying to transmute mercury into silver using an oil distilled from human feces.[51] Fontenelle remains steadfastly silent about Homberg's transmutational endeavors; indeed, he may have played a role in preventing Homberg's lifework from appearing before the public and reminding them of his transmutational claims.

Shortly after Homberg's death in 1715, Leibniz wrote a letter to discover the whereabouts of Homberg's papers. He received the answer that they were in the hands of Homberg's former student Étienne-François Geoffroy. These papers included two complete treatises ready for the press, one of which was the finished version of his otherwise incomplete *Essais de chimie* – a project on which Homberg had been laboring for nearly twenty years. Fontenelle himself mentions the complete manuscript of the *Essais* in his Éloge for Homberg.[52] According to Leibniz's informant, these manuscript treatises were to be, as Homberg desired, "published as soon as possible."[53] But nothing ever appeared. Given the "alchemical" origins of the experiments upon which so much of Homberg's *Essais* are built, and his unabashed claims successfully to have produced gold from mercury, a fuller version of the work – bearing Homberg's name and his title as academician – may well have been unwelcome.[54] (I should mention that Homberg's last paper published in the *Memoires*, in 1714, was an approach to the Helmontian alkahest following ideas propounded by George Starkey.[55]) This would not be the only example of censorship, since in 1680 the Académie denied permission to Homberg's chymical predecessor Samuel Cottereau Duclos to publish his own work on chymical principles, probably owing to his cosmological speculations and his pursuit of the alkahest and the Stone.[56]

Related to this possible suppression of Homberg's legacy and the general trend of the Académie to vilify and repudiate "alchemy" may well be the famous paper that Geoffroy published in 1722 entitled "Some cheats concerning the Philosophers' Stone." The paper relates methods used by supposed alchemists to trick the credulous into believing that they have witnessed a transmutation. But the majority of the paper is cribbed directly from the *Examen fucorum pseudochymicorum*, a well-known work published in 1617 by Michael Maier and intended to help fellow chrysopoeians distinguish true from false transmutations.[57] Thus Geoffroy's paper was nothing new, and one has to ask why an otherwise busy and productive chemist would bother to publish it at this date. Two related answers present themselves. One, Geoffroy was helping in the marginalization of transmutation and the concomitant elevation of the status of chemistry. Fontenelle took full advantage of Geoffroy's publication to write a lengthy précis in the *Histoire* containing some of his most vitriolic and sarcastic condemnations of "les Alchimistes." But I find it curious that Fontenelle also used this opportunity explicitly to distinguish alchemical claims to radically dissolve the metals from the decompositions of metals made with the burning lens by Homberg twenty years earlier in 1702 and by Geoffroy in 1709. Again Fontenelle seems to protest too much. He also asserts that alchemists have never made a single grain even of an imperfect metal.[58] This latter statement suggests a further incentive for the publication, one more personal for Geoffroy. Might this 1722 paper be a kind of a *shibboleth* from Homberg's student and the one who had claimed previously to have synthesized iron, only to be ridiculed by Louis Lemery for having drawn the experiment from Johann Joachim Becher, the "notorious" alchemist, and thus a disreputable source?[59] Since Geoffroy had been Homberg's closest associate, and since he himself quite probably supported chrysopoeia at least in his younger days, he had the greatest need to distance himself from Homberg and transmutation, given the official stance adopted by the Académie.[60]

But an enhanced status for chemists and chemistry came not only from denying their relationship with the potentially disreputable but also by creating relationships with the reputable. Thus Fontenelle's eloge of Homberg not only denies any interest on his part with *arcana*, but also provides him serial apprenticeships with more than a dozen notables of the late seventeenth century, even when it means that Fontenelle's chronology apprentices Homberg to people who would have been dead when Homberg met them.[61] Fontenelle is right enough to say that Homberg met Boyle, although my current research shows that it is unlikely he stayed with him to study, as Fontenelle claims, in "one of the most learned schools of physics."[62] Furthermore, I note with delicious irony, that the only thing I can confidently assert that Homberg *did* learn from Boyle was the secret preparation of Starkey's Philosophical Mercury for making the Philosophers' Stone.[63]

Of course, although my focus here has been on the expunging of the "alchemical" as a means by which the status of chemistry was elevated in this period, I do not intend to imply that this was the sole means. Contributors to this volume explicitly address other ways in which the discipline's status was enhanced throughout the eighteenth century – partly by entering university curricula, partly by public reeducation, and by acquiring ever-expanding utility and productivity for a host of applications.

During Homberg's formative years, that is, in the 1670s, there was no formalized place to go in order to obtain recognized credentials in chemistry, but that changed during the period under consideration here. The entry of chemistry in a formal and independent way into the university, and the training of students, for example around Stahl and Boerhaave, and down through many others later in the century is one of the crucial changes leading to the enhancement of the status of chemistry and chemists, and must make up an important part of any organizational narrative. Herman Boerhaave is a key link here. His 1718 academic oration "De chemia suos errores expurgante" cites the poor image of chemistry ("à commercio Sapientum remota, ignota Eruditis vel suspecta") I have just been outlining, as well as the "purging" the subject was then undergoing.[64] Boerhaave – and indeed, Leiden as a whole – likewise plays an important role in revamping chemistry, introducing it into university curricula, and arguing for its importance and utility.[65] John Powers' contribution to this volume details Boerhaave's recasting of chemistry in his lectures at Leiden. Rina Knoeff also treats the Leiden professor, addressing his view of the utility of chemistry for medicine.

Outside the university a new public image of chemistry had to be disseminated by popularization – correcting the deleterious notions the public had acquired from stage, rumor, and gossip – and so we must look also at lecturing and popularization throughout the century. Bernadette Bensaude-Vincent and Christine Lehman cover the rise and spread of public lecturing on chemistry in France. Besides augmenting educational possibilities in chemistry, these lectures clearly stressed chemistry's utility. The same twin issues – chemical education and utility – are covered by Ursula Klein in her consideration of apothecary-chemists in Germany. The utility theme appears again in an Scottish context in Matthew Eddy's study of James Anderson's application of chemistry to issues of agriculture and geology, and in an English context in Trevor Levere's account of Thomas Beddoes.

The last transformation of chemistry during the first half of the eighteenth century is the one that can be documented least well within the scope of this introductory statement, namely, the remarkable multitude and diversity of chymical theoretical systems proposed, developed, and rejected. This fertility of invention provides an outstanding forum for studying exactly how scientific theories emerge from laboratory results – particularly in a branch of science so closely tied to the laboratory and with such inherently complex phenomena as chemistry. It also provides a refutation of those who imagine that chymistry's development was largely "theory-free" (or at least "theory-poor"), driven primarily by artisanal practices and practical applications. Moreover, a consideration of this diversity drives home the danger of overly reductionist narratives which tend to overlook scenic routes and by-ways, no matter how heavily trafficked, to postulate some putative royal road for the development of the sciences.

Of particular importance for improving our understanding of the changes in chemistry during this period is the need to tease out, by means of careful and contextualized readings of experimental texts, the *implicit theories* lying behind specific experiments or operations. That is to say, our chymists often do not explicitly explain the thinking that led them to a particular experiment; however, it is essential if we are to understand the chymistry of the period that we do our best to analyze the ways in which they thought about their general work and their specific operations. Only thus can their mental landscape be revealed, and only thus can we uncover their formative ideas and implicit theories, and in turn reveal the crucial interplay between theory and experiment. Such study serves to overcome the residuals of the belief that before Lavoisier's theory and system, chymists labored in a state of confusion, or in ways that were "merely" empirical. Close reading and patient unpacking of specific experimental pathways can reveal that many late seventeenth- and early eighteenth-century chymists guided their work rationally, clear-mindedly, and systematically using the best resources at their disposal. Their theoretical frameworks might be neither ours nor Lavoisier's, but neither were their questions or their goals. All too often in the past, historians have confused our own lack of understanding of the context, goals, and implicit theoretical systems of earlier chymists with a lack of understanding on the part of those chymists themselves.

In the case of Homberg, a close analysis of his publications with attention to the specific choices he makes in experimental design reveals a dynamic chemical mind, carefully testing older theories and constantly developing, rejecting, and refining new ones. Here is one reason I emphasized the problem of casually labeling him a Cartesian – for such facile denominations tend to make us overlook the most interesting and revealing parts of the story. What follows here is only a single, isolated example.

Early in his career Homberg published a paper entitled "A method for extracting a volatile mineral acid salt in dry form." Two questions should immediately present themselves to the historian of science: why was he doing this? and, did he discover this method by chance or by design? Taking the paper on its own makes answering such questions nearly impossible, but by comparing this paper with Homberg's other contemporaneous work and unpublished archival sources, it becomes clear that he was involved in a larger project of theory-testing, namely, testing the Helmontian theory that all material substances are modifications of water.[66] The isolation of a dry acid salt from the well-known mineral acids would verify that they are in fact mostly water with only a relatively small quantity of dissolved volatile salt. Homberg would later go on to try to reduce the volatile acid salts themselves into water in an experiment that he patiently carried out over the course of four years. The isolation of this volatile acid salt would also silence some chemists to whom Homberg refers (without naming them) who denied the existence of volatile mineral salts. Thus already we can see an early eighteenth-century chemist deeply involved in theory-testing and theory-building – guiding his work with developed theoretical principles and conceptual structures, not just accumulating processes as some recent authors have suggested.

Homberg's practical difficulty was that it was impossible to separate the water from the salt by the usual method of distillation because the putative salt was as volatile as the water, and so they co-distilled. So, Homberg sought for an indirect method to separate the salt from the water. He begins with aqua fort (nitric acid), dissolves two ounces of silver in it, and then precipitates the solution with common salt, to obtain 2.5 ounces of silver precipitate. The weight increase of half an ounce indicates that the silver has successfully taken hold of the acid salt. Here we see Homberg's characteristic use of weights to monitor experiments – something which had been part of chymistry now for over a century, especially following the legacy of Van Helmont, and which for Homberg is now a routine and necessary tool.[67] But after the precipitation of the silver, the desired salt, although separated from the water, is now bound even more tightly to the silver – how does Homberg know this? Simply because "even hot water is not able to separate" the salt from the silver precipitate since washing it with hot water provokes no change. Now he tries to transfer the salt to another metallic substance from which it would be easier to separate. He first tries antimony, which he finds is in fact corroded by the silver precipitate. Since the silver is regenerated in its original form and weight, Homberg knows that the antimony has taken up all the salt from the silver. But alas, antimony proved as volatile as the salt, and so upon heating, both sublime together without separation – Homberg is no better off than when he started. Since the problem stems from antimony's volatility, Homberg decides to substitute a less volatile metal: tin. By gently heating the silver precipitate with tin filings, the silver is liberated (in its original weight), and a molten material composed of the tin combined with the acid salt can be poured off. Upon cooling, this liquid solidifies, and then by gentle heating Homberg can sublime away part of it as beautiful saline crystals. This sublimate weighs nearly half an ounce, is acidic, and dissolves readily in water and spirit of wine, demonstrating that it is the volatile acidic mineral salt that was initially in the aqua fort.

What do we learn from this close analysis? First, chymical experiments were designed and wielded *to test chymical theories*, such as Van Helmont's, in the late seventeenth and early eighteenth century. Second, Homberg clearly had in mind the idea of permanent or semi-permanent chemical species that survive reactions. The acid salt was not destroyed by the dissolution of the silver, rather it was incorporated with the silver in a recoverable form; likewise, the silver acted only as a temporary carrier of the salt, and was recovered unchanged. This, by the way, is quite different from Lemery, who claims that the points of acid particles are broken in the course of a reaction, and thus one trying to separate them again in their original form "would never succeed."[68] Third, Homberg had in mind both a treasury of factual information about the discrete properties of substances as well as operational principles very close to the ideas later incorporated formally into Geoffroy's famous *Table des rapports*.[69] That is, he implicitly uses a series of relative affinities to transfer the acid salt: first from water to silver, then from silver to tin, and then away from the tin by means of heat. Running alongside the theory is a repertory of manual operations – solution, precipitation, sublimation, and so forth – that are necessary for actually carrying out the designed experiment practically. Homberg's volatile acid salt experiment – and this is only one of many which do so – indicates a serious level of experimental design based on systematic theoretical principles in order to carry out a specific operational or preparative goal. This level of theoretical engagement has not always been adequately represented in our accounts of the chymistry of the period, partly because there has been a dearth of careful, informed analyses of the details of chymical experiments. The point is that a careful unpacking of implicit theoretical systems can inform us more fully about the thought-processes of early eighteenth-century chymists than do their often-scanty explicit theoretical statements.

It remains to be illustrated how a careful unpacking of implicit theoretical frameworks will alter our understanding of the filiation and development of theories of chemistry in the first half of the eighteenth century. Such an investigation may well indicate not only hitherto obscured connections between different chemists or chemical centers, but also a much richer, more diverse set of chemical schools of thought than was perceptible under the conditions of earlier historiographical narratives that privileged a very small number of lineages at the expense of others. In any event, it seems to me that we will undeniably attain a clearer and more accurate picture of what chymists were actually thinking by carefully examining the specifics of what they were doing in their laboratories and how they chose what courses to pursue.

While emphasizing the importance and diversity of chymical theories in the period, I need however equally to emphasize the peculiar status of theories in chemistry as I try to draw together some of the diverse points of this paper. Chemistry has always been about the physical, the sensual, and the productive. Chymists, whether medieval seekers for transmutation or eighteenth-century academicians (and even most chemists of the twenty-first century) want to *make stuff* – new materials with specific properties. Thus chemical theories tend to be valued at least as much for the guidance they promise in practical, productive endeavors as for the understanding of the world they might offer. Practicing chymists of the early eighteenth-century often simply did not see any practical value in some of the more abstract theoretical frameworks – like the mechanical philosophy – propounded by more philosophical or systematic wits. Therefore, such grand systems that loom so large in our views of the period often washed over chymists with little apparent effect.[70] Chemistry thus occupies a unique place among the natural sciences – where theory and practice are unusually closely linked, and where experiment has a meaning and goal quite distinct from the largely probatory and exploratory role set out for it in traditional accounts of experiment.

This link to material production of course accounts for some of the low status of chymistry discussed earlier. Links to artisanal, productive, and commercial practices possibly counted against chymistry as much as rumors of cheating pseudo-chrysopoeians. The same is true of the widespread unwillingness of most of the chymists in our period to philosophize in purely cerebral ways. Homberg, whose theory was the best developed of all early eighteenth-century systems, always kept laboratory processes squarely before his eyes as the *primary* path to understanding. Indeed, Nicolas Remond, a great fan of Descartes and Leibniz and one of Homberg's associates in the household of the Duc d'Orléans, remarked that "according to [Homberg], all philosophy lay in the use of the fire-tongs, and thus he cared little for either the ancients or the moderns."[71] While the sentiment was probably intended as a bit of an insult, Homberg would probably have accepted it with a measure of pride. Fontenelle, who plays such a large role in this story, expresses a similarly dismissive attitude, for he seems always impatient for more "philosophical" principles in chemistry – something akin to the axioms of mathematics or physics. Chemistry had to be philosophically domesticated and civilized – for Fontenelle, this meant aligning it with the "esprit géométrique" of Descartes; for later historians of chemistry, it too often meant aligning it with "greater" currents in the history of ideas.

The history of science long privileged thinking over doing, understanding over producing, theory over practice. This inclination undoubtedly encouraged the linking of chemical developments to those great thinkers and systematizers Descartes and Newton. Thus it seems that some earlier historians of chemistry were, consciously or not, continuing the campaign to elevate the status of their subject by linking it to acknowledged and cerebral "greats" rather than to grimy laboratories. But this practice has had the effect of diminishing the very features that are the most characteristic of chymistry. Any new organizing narrative proposed today must take into full account the ever-present role of material practice and experiment within chymistry.

When narratives of the development of chymistry point to tides and winds of changing philosophical fashions (for example, from Paracelsian to Cartesian to Newtonian) as motive forces, they imply that chymical experiments and observations are infinitely flexible, able to be bent, twisted, and reinterpreted in any direction to fit some "larger" agenda or system. External theories, whether from physics or philosophy, are made to trump chymical practices and observations. Thus, the important interplay between experimental results and theoretical frameworks in the mind of the chymist is deemphasized or neglected. Of course, no serious historian today would question the influence of philosophical and theoretical systems on the way experimental results are interpreted. However, there is serious danger in undervaluing the transformative power of gritty experimental results. To do so renders the entire material practice of chymistry – in short chymistry's most essential feature – relatively unimportant and without firm epistemic status, and can reduce the chymist himself to a cipher. Such a scenario ignores the way most of our historical characters actually spent their time: in laboratories, getting dirty and trying to make sense of a dizzying array of phenomena and results, constantly tinkering – in rational and historically analyzable ways – with both theoretical principles and chymical operations.

Current historians of science should not blush if the developments in chymistry of the period came not from the exalted philosophizings and systematizings of Descartes or Newton, but rather from the crass sexual metaphors of Tachenius, the sooty fire-tongs of Homberg, and the dung-baths and putrefying urine of countless others. I am confident that we will learn a great deal more about chymistry when it is taken on its own terms rather than on borrowed ones, and that will require close attention to chymistry's technical aspects and the always close interplay between theories and practices. When Boyle referred to "the Chymists" as "sooty Empiricks," he was only half right. The chymists *were* sooty, and happily so, but they were not on that account "empirics." Recent work shows clearly how theory and practice were in constant dialogue in the work even of aspiring chrysopoeians, and even back to the medieval foundations of chymistry; this sort of engagement needs to be recognized, and its features delineated for a much wider array of chymists.[72] Accounts of the history of chemistry should not recoil at the soot or at the fiery labors and thought processes that accompanied its production.

Finally, it seems to me that when properly carried out, the history of chemistry promises to make unique contributions to the history of science in general and to the literature on experiment and the relations between science and technology in particular. We miss an important feature of chemistry if we adopt, implicitly or explicitly, an historical narrative that over-emphasizes the role of gaining understanding at the expense of gaining productive power over the natural world. The uniqueness of chemistry and the history of chemistry thoughtfully done have much to offer the history of science more broadly.

NOTES

[1] Seymour Mauskopf's epilogue to this volume provides an excellent overview of the rise and reign of the Lavoisier-centered narrative.

[2] Robert Siegfried, *From Elements to Atoms: A History of Chemical Composition, Transactions of the American Philosophical Society* 92, part 4, 2002, 56–73 ("Stagnation of Chemical Theory: 1675–1750").

[3] In using the archaically spelt word chymistry, I include both what we call "chemistry" and "alchemy" under a single term used at the time in which the two were not separated. This usage follows the recommendations that both I and William Newman set forth independently and jointly some time ago, see for example, William R. Newman and Lawrence M. Principe, "Alchemy vs. Chemistry: The Etymological Origins of a Historiographic Mistake," *Early Science and Medicine* 3, 1998, 32–65, on 59–61.

[4] One example of this in regard to the subject of alchemy is Lawrence M. Principe and William R. Newman, "Some Problems in the Historiography of Alchemy," 385–434, in *Secrets of Nature: Astrology and Alchemy in Early Modern Europe*, eds. Newman and Anthony Grafton, (Cambridge, MA: MIT Press, 2001).

[5] Hélène Metzger, *Les doctrines chimiques en France du début du XVIIe à la fin du XVIIIe siècle* (Paris: Les Presses Universitaires, 1923).

[6] Hélène Metzger, *Newton, Stahl, Boerhaave et la doctrine chimique* (Paris: Librairie Scientifique Albert Blanchard, 1930).

[7] Pierre Brunet, *L'introduction des théories de Newton en France au XVIIIe siècle* (Paris: Librairie Scientifique Albert Blanchard, 1931).

[8] Henry Guerlac, *Newton on the Continent* (Ithaca, NY: Cornell University Press, 1981).

[9] Marie Boas [Hall], *Robert Boyle and Seventeenth-Century Chemistry* (Cambridge: Cambridge University Press, 1958); A. R. Hall, *The Scientific Revolution 1500–1800* (Boston: Beacon Press, 1966), 327–28; Arnold Thackray, *Atoms and Powers: An Essay on Newtonian Matter Theory and the Development of Chemistry* (Cambridge, MA: Harvard University Press, 1970), esp. 83–123; Mi Gyung Kim, *Affinity, That Elusive Dream* (Cambridge, MA: MIT Press, 2003), e.g. 11–13, 83.

[10] For example, Richard S. Westfall, *The Construction of Modern Science* (Cambridge: Cambridge University Press, 1971), 68–73. It must of course be noted that there was already in seventeenth-century France a spectrum of "Cartesianisms," yet Lemery's system cannot be included among them in any meaningful sense; on the former point see Tad Schmaltz, *Radical Cartesianism: The French Reception of Descartes* (Cambridge: Cambridge University Press, 2002), 9–12, and 19.

[11] The chymical part of Descartes' work is found predominantly in *Principes de la Philosophie* (1647). See Bernard Joly, "Descartes et la chimie," 216–21 (CD) in *L'esprit cartésien*, eds. B. Bourgeois and J. Havet (Paris: Vrin, 2000).

[12] On Tachenius, see James L. Partington, *A History of Chemistry*, 4 vols. (London: Macmillian, 1961), 2:291–96; Heinz-Herbert Take, *Otto Tachenius, 1610–1680: Ein Wegbereiter der Chemie zwischen Herford und Venedig* (Bielefeld: Verlag für Regionalgeschichte, 2002).

[13] Otto Tachenius, *Hippocrates Chymicus* (London, 1677), 87 and 89; *Clavis*, (London, 1677) 13.

[14] Ibid., 18; for more on Tachenius' sexual metaphors and language, see Principe, "Revealing Analogies: The Descriptive and Deceptive Roles of Sexuality and Gender in Latin Alchemy," in *Hidden Intercourse: Eros and Sexuality in Western Esotericism*, eds. Wouter Hanegraaff and Jeff Kripal (Chicago: University of Chicago Press, forthcoming, 2007).

[15] Marie Boas Hall, "Acid and Alkali in Seventeenth Century Chemistry," *Archives internationales d'histoire des sciences* 9, 1956, 13–28, on 15–16; Michel Bougard, *La chimie de Nicolas Lemery* (Turnhout: Brepols, 1999), 190–96; Bougard also highlights the role of Pierre Gassendi's particulate system as a background to Lemery's thought.

[16] In 1672, 1677, and 1680; an English edition appeared in 1689. On the link of St. André with Lemery, and supporting arguments that Lemery's chymical doctrine "n'est pas cartesienne" see Bernard Joly, "L'anti-Newtonianisme dans la chimie française au début du XVIIIe siècle," *Archives internationales d'histoire des sciences* 53, 2003, 213–24, esp. 215–16.

[17] For example, Kim, *Affinity*, 6, 12, and passim.

[18] This rejection occurred in the context of a dispute carried out in 1700–01 between Homberg and Johann Bernoulli on the luminescence of mercury enclosed in barometers.

[19] On the definition of "Cartesian" in this period, see the useful proposals and definitions made in Bernard Marsak, *Bernard de Fontenelle: The Idea of Science in the French Enlightenment* (Philadelphia: American Philosophical Society, 1959), and Tad M. Schmaltz, *Radical Cartesianisms: The French Reception of Descartes* (Cambridge: Cambridge University Press, 2002), 9–24. On Louis Lemery, see Bernard Joly, "Quarrels between Etienne-François Geoffroy and Louis Lemery," 203–14 in *Chymists and Chymistry*, ed. Lawrence M. Principe (Canton, MA.: Science History Publications/CHF, 2007).

[20] Étienne-François Geoffroy, "Des différents rapports observés en chymie entre differentes substances," *Mémoires de l'Académie Royale des Sciences* (hereinafter *MARS*) 20, 1718, 202–12; Guerlac, *Newton*, 77; William A. Smeaton, "E. F. Geoffroy Was Not a Newtonian Chemist," *Ambix* 18, 1971, 212–14; Frederic L. Holmes, "The Communal Context for Etienne-François Geoffroy's 'Table des rapports'," *Science in Context* 9, 1996, 289–311. It is true, however, that Torbern Bergmann (most notably) applied Newtonian ideas to his own affinity theories later in the century, but such a linkage dating from later in the century cannot be read backwards to Geoffroy and other early versions of affinity.

[21] Jean-Baptiste Sénac, *Nouveau cours de chymie, suivant les principes de Newton & de Sthall* (Paris, 1723), liii–liv, 26, 74–78. Note that the book was published anonymously, and the attribution to Sénac, otherwise known only as a physician, is open to question.

[22] John Freind, *Praelectiones chymicae* (Oxford, 1709); [Gottfried Wilhelm Leibniz], *Acta eruditorum*, 1710, 412–16.

[23] Lawrence M. Principe, "Wilhelm Homberg's *Wanderjahre*: Intellectual Formation and Transnational Networks," paper presented at "Science in Europe, Europe in Science" a conference held at Maastricht, 4–6 November 2004; Principe, "Wilhelm Homberg and the Chymistry of Light," paper presented at CalTech, 22 February 2004. This material will be published in my forthcoming book, *Wilhelm Homberg and the Transmutations of Chymistry at the Académie Royale des Sciences*.

[24] E.g. Richard S. Westfall, "Newton and Alchemy," 315–35 in *Occult and Scientific Mentalities in the Renaissance*, ed. Brian Vickers (Cambridge, Cambridge University Press, 1984) and *Never at Rest: A Biography of Isaac Newton* (Cambridge: Cambridge University Press, 1980), 299–308, 527–29.

[25] John Henry, "Occult Qualities and the Experimental Philosophy: Active Principles in Pre-Newtonian Matter Theory," *History of Science* 24, 1986, 355–81; William R. Newman, "Newton," in *New Dictionary of Scientific Biography* (New York: Scribners, 2007).

[26] For a similar conclusion using other data, see Joly, "L'anti-Newtonianisme dans la chimie française."

[27] A full treatment of Homberg and his work appears in my forthcoming book, *Wilhelm Homberg and the Transmutations of Chymistry*; Holmes eloquently argued for the importance of both the Académie and Homberg in his *Eighteenth-Century Chemistry as an Investigative Enterprise* (Berkeley, CA: Office for History of Science & Technology, University of California at Berkeley, 1989), and "Communal Context."

[28] Lawrence M. Principe, "Evidence for Transmutation in Seventeenth Century Alchemy," 151–64, in *Scientific Evidence: Philosophical Theories and Applications*, ed. Peter Achinstein (Baltimore: Johns Hopkins Press, 2005); on the acceptability of chrysopoetic endeavors for intellectuals, see Principe, "D. G. Morhof's Analysis and Defence of Transmutational Alchemy," 138–53 in *Mapping the World of Learning: The Polyhistor of Daniel Georg Morhof*, Wolfenbüttler Forschungen 91 (Wiesbaden: Harrassowitz, 2000).

[29] Lawrence M. Principe, "Wilhelm Homberg: Chymical Corpuscularianism and Chrysopoeia in the Early Eighteenth Century," 535–56 in *Late Medieval and Early Modern Corpuscular Matter Theories*, eds. C. Lüthy, J. E. Murdoch, and W. R. Newman (Leiden: Brill, 2001).

[30] Wilhelm Homberg, Voenno-meditsinskoi Akademii, Boerhaave Archive, MS 130, fols. 233r–v; "J'ay… poursuivre tout l'ouvrage de Philalethe."

[31] Ibid., fols. 119v–27r.

[32] Niedersächsische Landesbibliothek Hannover, Leibniz Briefe 420, fols. 3r–v; Leibniz to Homberg, 10 March 1711.

[33] Wellcome Institute Library, MS 2298, "Essai pour développer la science & la practique de l'Oeuvre des Philosophes chimiques," 5–7, 46–49, 84. Another copy is Université de Bordeaux, MS 23.

[34] Nicolas Lemery, *Cours de chymie*, 3rd edition (Paris, 1679), 57–60; see also Newman and Principe, "Alchemy vs. Chemistry," 59–61.

[35] *Oeuvres diverses de M. de Fontenelle*, 3 vols. (Paris, 1724), 1:1–35 (not paginated), "Sur l'utilité des mathematiques et de la physique," on sig. Aiiiiv.

[36] Ibid., 1:117–20.

[37] Wilhelm Homberg, "Essais de chimie," *MARS* 4, 1702, 33–52 on 33; Rémi Franckowiak and Luc Peterschmitt, "La chimie de Homberg: Une vérité certaine dans une physique contestable," *Early Science and Medicine* 10, 2005, 65–90; Homberg, MS 130, fol. 112v.

[38] Fontenelle, "Èloge de M. Lemery," *Histoire de l'Académie Royale des Sciences* (hereinafter *HARS*) 17, 1715, 75–76; on the textbook tradition see Metzger, *Doctrines*; on Lemery see Michel Bougard, *Chimie de Lemery*.

[39] Bourdelin, Bibliothéque Nationale, MS n. a. fr. 5148, fol. 1v (24 March 1699); "toutes ses experiences ne furent point trouvéez nouvelles. Mr. Hombert les avoit faites la pluspart."

[40] Principe, "Chymical Corpuscularianism," 538.

[41] Michel Chillat(?), *Les Souffleurs* (Paris, 1695): "Que la Chimie est admirable," 99–100. On the play see also Didier Kahn, "L'alchimie sur le scène française aux XVIe et XVII siècles," *Chrysopoeia* 2, fasc. 1, 1988, 62–96. It is reported by Maupoint, *Bibliothéque des théâtres* (Paris, 1733), 288, that the play was not performed; even if this report is true, the work was very popular as witnessed by the five printed editions that appeared within 18 months.

[42] *Archives de la Bastille*, 19 vols. (Paris, 1866–1904), esp. vol. 12 (1881): *Règnes de Louis XIV et de Louis XV* (1709 à 1772), 1–5, 52–68; Clara de Milt, "Christophle Glaser," *Journal of Chemical Education* 19, 1942, 53–60; Arlette Lebigre, *1679–1682, L'Affaire des poisons* (Brussels: Complexe, 2001).

[43] Louis de Rouvroy, duc de Saint-Simon, *Mémoires*, ed. Yves Coirault, 8 vols. (Paris: Gallimard, 1983–88), 4:459–66; *Aus der Briefe der Herzogin Elisabeth Charlotte von Orléans an die Kurfürstin Sophie von Hannover*, ed. Eduard Bodemann, 2 vols. (Hannover, 1891), 2:302–303, 307.

[44] Christoph Meinel, "Theory or Practice? The Eighteenth Century Debate on the Scientific Status of Chemistry," *Ambix* 30, 1983, 121–32.

[45] Newman and Principe, "Alchemy vs. Chemistry." See also John C. Powers, "'Ars sine arte': Nicholas Lemery and the End of Alchemy in Eighteenth-Century France," *Ambix* 45, 1998, 163–89.

[46] An excellent and sensitive study of the subject of fraud in alchemy is Tara E. Nummedal, *The Battle for Alchemical Authority in the Holy Roman Empire* (Chicago: University of Chicago Press, 2007).

[47] For analogous comments at the Royal Society, see Thomas Sprat, *History of the Royal Society* (London, 1667), 37–38; recall that the Royal Society's most prominent Fellow, Robert Boyle, was simultaneously pursuing traditional chrysopoeia.

[48] Jean-Baptiste Duhamel, *Regiae scientiarum academiae historia* (Paris, 1698; enlarged edition, 1701).

[49] Fontenelle, "Éloge de M. Lemery," *HARS* 17, 1715, 73–82 on 73.

[50] Fontenelle, "Éloge de M. Homberg," *HARS* 17, 1715, 82–93 on 87–88.

[51] Wilhelm Homberg, "Observations sur la matiere fecale," *MARS* 13, 1711, 39–47.

[52] Fontenelle, "Homberg," 92. I have recently, after lengthy searching and negotiation, recovered a copy of this long-lost and complete version of Homberg's *Essais*; full details are forthcoming in Principe, *Wilhelm Homberg*.

[53] NLB, Leibniz Briefe 768, fols. 53r–54v, on fol. 54r; Nicolas Remond to Leibniz, 23 December 1715.

[54] Principe, "Chymical Corpuscularianism."

[55] Wilhelm Homberg, "Mémoire touchant la volatilisation des sels fixes des plantes," *MARS* 16, 1714, 186–95.

[56] Alice Stroup, "Censure ou querelles savantes: L'Affaire Duclos (1666–1685)," in *Règlement, usages et science dans la France de l'absolutisme*, eds. Christiane Demeulenaere-Douyère and Éric Brian (Paris: Lavoisier Tec et Doc, 2002), 435–52.

[57] Étienne-François Geoffroy, "Des supercheries concernant la pierre philosophale," *MARS* 24, 1722, 61–70; Michael Maier, *Examen fucorum pseudo-chymicorum detectorum et in gratiam veritatis amantium succincte refutatorum* (Frankfurt, 1617); Wolfgang Beck, *Michael Maiers Examen Fucorum Pseudo-chymicorum: eine Schrift wider die falschen Alchemisten*, Ph.D. 1992, Technische Universität München. Robert Halleux, "L'alchimiste et l'essayeur," in *Die Alchemie in der europaeischen Kultur- und Wissenschaftsgeschichte*, ed. Christoph Meinel (Wiesbaden: Otto Harrasowitz, 1986). It is to be noted that much of Maier's work is in turn borrowed from Heinrich Khunrath, *Trewhertzige Warnungs-Vermahnung* (Magdeburg, 1597).

[58] Fontenelle, *HARS* 24, 1722, 37–39.

[59] Louis Lemery, "Nouvel éclarcissement sur la prétendüe production artificielle du fer, publiée par Becher & soûtenüe par M. Geoffroy," *MARS* 10, 1708, 371–402. A brief account of this debate is given in Metzger, *Doctrines*, 407–09; a full account appears in Joly, "Quarrels."

[60] Support for Geoffroy's own interest in chrysopoeia comes from his library, which contained more than seventy books on the topic, including classic works by Philalethes, Valentine, and others, as well as Manget's huge 1702 compendium of chrysopoetic texts, *Bibliotheca chemica curiosa*; see *Catalogus librorum Stephani-Francisci Geoffroy* (Paris, 1731). Moreover, it is interesting to note that Geoffroy does not actually use his 1722 paper to debunk chrysopoeia itself, but rather, as Maier did, simply to point out the likelihood of fraud; it is Fontenelle who uses Geoffroy's paper as a jumping-off point for a full-scale assault against transmutation itself.

[61] Alice Stroup, "Wilhelm Homberg and the Search for the Constituents of Plants at the 17th-Century Académie Royale des Sciences," *Ambix* 26, 1979, 184–202, on 185–86.

[62] Fontenelle, "Homberg," 85.

[63] Principe, *Wilhelm Homberg*, chapter 1; "Chymical Corpuscularianism," 546–47.

[64] Herman Boerhaave, "Sermo academicus de chemia suos errores expurgante," published in *Elementa chemiae*, 2 vols. (Paris, 1733), 2:64–77, on 65.

[65] See John C. Powers, "Chemistry Enters the University: Herman Boerhaave and the Reform of the Chemical Arts," *History of Universities* 21, 2006, 77–116.

[66] Wilhelm Homberg, "Maniere d'extraire un sel volatile acide minéral en forme séche," *Histoire et Mémoires de l'Académie Royale des Sciences 1666–99*, 11 vols. (Paris, 1729–33), 10:202–08 (paper read on 31 December 1692); Principe, *Wilhelm Homberg*, chapter 3; on Helmont's water theory see Walter Pagel, *Joan Baptista van Helmont* (Cambridge: Cambridge University Press, 1982), esp. 49–60.

[67] William R. Newman and Lawrence M. Principe, *Alchemy Tried in the Fire: Starkey, Boyle, and the Fate of Helmontian Chymistry* (Chicago: University of Chicago Press, 2002), 46–91.

[68] Nicolas Lemery, *Cours de chymie* (Paris, 1675), 10.

[69] Geoffroy, "Rapports observés."

[70] This fact helps explain the rather minor impact on the content and practice of chemistry made by that theory's greatest exponent, Robert Boyle; see Principe, "Les liens chymiques entre Boyle et France," forthcoming in *Robert Boyle et la philosophie naturelle*, eds. Charles Ramond and Miriam Dennehy (Paris: Vrin, 2007).

[71] NLB, Leibniz Briefe 468, fols. 53r–54v; Remond to Leibniz, 23 December 1715; on fol. 54r: "toute la philosophie selon lui etoit dans l'usage de la pincette et ainsi il faisoit peu de cas des anciens et des modernes."

[72] Newman and Principe, *Alchemy Tried in the Fire*, esp. 92–155.

KEVIN CHANG

GEORG ERNST STAHL'S ALCHEMICAL PUBLICATIONS: ANACHRONISM, READING MARKET, AND A SCIENTIFIC LINEAGE REDEFINED

This paper examines the development of Georg Ernst Stahl's alchemical thinking and an apparent anachronism that the complexities of Stahlian publications introduced. Stahl's development involves his shift from the position of an alchemical believer or aspirant to that of a disbeliever openly critical of alchemy, and the transformation of his relationship with Johann Joachim Becher, an important chemist of early modern Europe. It also involves an examination of the publishers' interests that caused an apparent anachronism, and the practices of publication at the time, including the introduction of academic works to the German-reading public.

Stahl's prominence in eighteenth-century chemistry seems to be beyond doubt. One can easily enumerate a remarkable list of important chemists who either identified themselves as Stahlians or who were identified as such in the second half of the eighteenth century in Europe: Guillaume François Rouelle, Pierre-Joseph Macquer, Antoine Baumé, G. F. Venel, Joseph Priestley, Henry Cavendish, Joseph Black, Richard Kirwan, Carl W. Scheele, and so on.[1] This list is sufficient to testify to Stahl's enormous influence. Admittedly, Stahl's influence is usually portrayed rather negatively in the predominant historiography of eighteenth-century chemistry, since his phlogiston theory is said to have formed a formidable obstacle to the modern chemistry that Antoine Lavoisier championed. This historiography suggests that during the last quarter of the eighteenth century, Lavoisier and his comrades fought a difficult but heroic battle that overcame Stahlian chemistry and laid the foundations for both the modern understanding of combustion and the system of chemical elements.

This historiography commits an anachronism of its own, for it takes for granted that one can identify the phlogiston theories of late eighteenth-century chemists as simply "Stahlian," and thus hardly goes further to inquire into the sources and trajectories of these later Stahlian theories. These trajectories certainly led to the eve of the so-called Chemical Revolution championed by Lavoisier, and the historiography seems to suggest that the last third of the eighteenth century was the heyday of Stahl's influence, especially in France and England. But if we look back over the course of the eighteenth century, we soon come to realize that we know little about the development of Stahlian chemistry from his lifetime through the second third of the eighteenth century. Thus, given Stahl's great influence, we actually have a very limited knowledge of where and how it came to be. Further, Stahl's *influence*

L. M. Principe (ed.), New Narratives in Eighteenth-Century Chemistry: Contributions from the First Francis Bacon Workshop, 21–23 April 2005, 23–43.

must be distinguished from Stahl's *own thought*. For example, we are cautioned that, although the Stahlian chemistry that Rouelle set forth in mid-century Paris was indeed inspired by Stahl's work, Rouelle's chemistry did not in fact follow Stahl's teaching faithfully.[2] The historiography of chemistry too often understands Stahl's work retrospectively, seen through the view of late eighteenth-century Stahlians, or even through that of their opponents, yet neither of these views is necessarily faithful to Stahl's own chemical teaching. To differentiate one from the other requires at least knowledge of Stahl's own work, and yet there remains little primary work on Stahl's chemical teaching that is reliable. This insufficiency of primary work on Stahl permits the emergence of the apparent anachronism I will discuss.

This paper hopes to rectify an inadequacy in Stahl studies by examining closely the transformation of Stahl's views on alchemy. Recent writers on Stahl's alchemy have presented two conflicting pictures. Some historians highlight Stahl's belief in alchemy, while others, such as Karl Hufbauer and Johanna Geyer-Kordesch, portray Stahl as a resolute critic of alchemy.[3] Curiously, *both* of these pictures are true to the texts that these scholars examine; the problem is that neither addresses the *transformation* of Stahl's position. J. R. Partington, whose chapter on Stahl is one of the most important twentieth-century secondary sources, notes only briefly that Stahl's position shifted – a surprising fact since we know that he always paid keen interest to alchemy in his *History of Chemistry*.[4] It seems that the only historian who has closely examined the transformation of Stahl's views is Hermann Kopp, who, in a mere seven pages in *Die Alchemie bis zum Letzten Viertel des 18. Jahrhundert*, provided an admirable survey of Stahl's discussions on alchemy.[5] Yet even Kopp did not point out the curious anachronism that Stahl's publications on alchemy might imply, and in particular he did not elaborate on the transformation of Stahl's intellectual relationship to Becher.

This paper will elucidate this apparent anachronism in regard to Stahl's views on alchemy in the context of shifting ideas and relationships by investigating the course of Stahl's alchemical publications and the works by Becher that Stahl either helped to republish or for which he wrote prefaces. First, we will analyze and reproduce at some length Stahl's arguments for and against chrysopoeia, a term that Stahl used to specify gold-making, thus coinciding with William R. Newman's and Lawrence M. Principe's recent terminological proposal.[6] The substantial reproduction of Stahl's supportive and critical arguments can serve as a foundation for precise discussions on both Stahl's teachings and his place in the history of chemistry. Second, this study will take into account the role of publication in mediating an author's ideas. We will focus on three of Stahl publications on chrysopoeia: two published in 1720 with the titles *Chymia rationalis et experimentalis*[7] and *Gedancken von Verbesserung der Metallen*,[8] and a third, "Bedencken von der Gold-Macherey," published in 1726 as a foreword to the republication of Becher's *Chemischer Glücks-hafen*,[9] a collection of chymical recipes.[10] Our understanding of Stahl's connection to alchemy is complicated by two entrepreneurial figures in the world of publication, each of whom published one of the two 1720 alchemical texts by Stahl. One of them, Friedrich Roth-Scholtz, is well known in the alchemical literature for publishing the famous *Bibliotheca Chemica* and *Deutsches Theatrum Chemicum*. The other, Caspar Jacob Eyssel, is less well known in the history of publication but was important in popularizing Stahl's academic works among a German-reading market. Thus at the heart of this case study are not only the ideas of a scientist, but also issues that have not received much attention in the historiography of the history of chemistry, namely, the relationships between a text and its author, and between an author and his publisher, the latter driven by his own publication agenda and the market value of the author and his work.

1720: THE *CHYMIA RATIONALIS* AND ITS PUBLISHER

The year 1720 was an eventful one for Stahl's alchemical publications; it saw the publication of two of Stahl's texts relating to his alchemical thinking in previous years. Stahl's *Chymia rationalis et experimentalis* was a textbook compiled from notes taken by students who attended Stahl's lecture course in chemistry at the University of Jena in 1684, the year when he received his doctorate. Therefore, it reflects his teaching at the very outset of his professional career. The other publication of 1720, *Bedancken von Verbesserung der Metallen*, put into the press by Roth-Scholtz, was the German translation of a Latin text on the legitimacy of chrysopoeia first published in 1703.[11] These two works, both speaking affirmatively of chrysopoeia, recorded Stahl's teaching at two different points earlier in his career, and were published at a time when Stahl had secured his position as First Physician to a powerful king, the so-called Soldier King, Friedrich Wilhelm I of Brandenburg-Prussia, and had become the highest official in chemico-medical matters in that kingdom.[12] Eighteenth-century readers, as well as those today, may simply assume that these texts were true to the thinking of this established chemico-medical author at the time of their publication; apparently that is what their publishers wished readers to believe.

The *Chymia rationalis*, a German text in the format of a textbook, was Stahl's best-known chemical work in the English-speaking world, since it was closely related to the only English translation of Stahl's work, *Philosophical Principles of Universal Chemistry*, published by Peter Shaw in 1730.[13] Both the German text and Shaw's English translation are said to be based on a Latin version, although the first Latin edition, *Fundamenta chymiae*, was not published until three years after the German version.[14] While the foreword of the 1720 German edition differs from that of the later Latin one, the bodies of the two editions are essentially identical. Since Stahl himself, his publishers, and his bibliographer never touched upon the possibility that the German translation relied on a different manuscript, and since their contents are almost identical, it is reasonable to assume that the German edition was in fact a translation of a manuscript that was eventually published as the Latin edition.

What is important to note is that the Latin manuscript was not Stahl's own composition, and that it circulated beyond Stahl's reach. There are clues to suggest that this manuscript was in circulation before 1720. First, the foreword to the Latin edition is dated 10 December 1720 by "J. S. C.," generally construed to be Johann Samuel Carl, a former student of Stahl's who later rose to be a chemico-medical figure of some importance. Second, the foreword to the German edition identified its own author as a *translator* (*Übersetzer*), thus confirming that the Latin original was already available by 1720. Indeed, this translator indicated in his foreword that Stahl's lecture notes had previously existed in the form of a manuscript in the hands of former students and a few *Liebhabern* that loved this subject.[15] In fact, the manuscript was so much sought after that interested readers paid to make scribal copies from it. Its clear commercial potential thus obviously motivated the publishers to introduce it to the market of printed books. Although we do not know exactly why the German translation came out before the Latin original, evidently its commercial value was so appealing that the German and Latin editions appeared within three years of each other.

The alchemical nature of the body of the textbook is not as obvious as that of its two supplements. The title page indicates that the first supplement is a treatise studying the mercuries of metals, animated mercury, and the philosophers' stone. It also indicates that the other supplement is a tract by Isaac Hollandus, a legendary fifteenth-century adept of alchemy, entitled *Issaci Hollands Tractat von den Salzen und Oehlen der Metallen* (*On the Salts and Oils of Metals*). While the forewords of the German and Latin editions indicate nothing explicitly about the authorship of the first supplement, their silence simply implies that Stahl was its author. Although Stahl would express regret over the publication of this text, he never denied that the textbook with its two supplements reflected his teaching at Jena in 1684.

Both of the supplements to *Chymia rationalis* deal with transmutational operations. The first examines various processes of mercurification, that is, the rendering of metals into their "mercuries," their supposed primordial states with a liquidity and luster resembling quicksilver. If properly prepared, these mercuries could be turned into the "philosophical mercury" or animated mercury, which is required as the starting material of transmutation. The author compares three schools of chrysopoeia that differ in their chief chemical instruments for preparing the philosophers' stone (also identified as the "tincture") and thereafter proposes the substances necessary for transmutation as well as the means whereby the philosophers' stone is prepared.[16] The tract by Hollandus likewise deals with methods of achieving transmutation by working from the salts and oils of metals. Even though the latter appendix is not Stahl's own composition, the substance and tone of Stahl's own study of mercurification and the philosophers' stone leaves no doubt that he is a believer in traditional alchemical transmutation.

The publisher of the German edition, the Leipzig-based Caspar Jacob Eyssel, announced in the foreword that the publication of this chymical textbook was just one in a series of publications of Stahl's works upon which he was embarking.[17] The move to publish Stahl's works started in 1714 – just a year before the professor of medicine moved to Berlin – with the German translation of a text on clinical observations that was also compiled from lecture notes taken by one of his students.[18] Apparently, this publication was initiated by the student who compiled the notes, for the name of the student, Gottfried Heinrich Ulau, appeared on the title page, while the instructor – Stahl – was referred to anonymously simply as a "very experienced man" (*experientissimus vir*). In 1716, Eyssel published a German translation of Stahl's work on spas and hot springs, again leaving the author anonymous, but this time referring to him as "a famous physician in Berlin."[19] His first Stahl publication on clinical observations was so successful that in 1718 a second edition was already up for sale. This edition was given a new title that emphasized Stahl's name and position at court: "Georg Ernst Stahl, Famous Royal *Leib-Medicus* of Prussia, and Court Counselor," a sign that suggests in part that Stahl's status at court in Berlin was secure and considered helpful for the sales of the book.[20] Eyssel sped up his Stahl translation project by publishing two of his medical dissertations in 1718, one in 1719, and one more in 1720,[21] as well as a German translation of his most important metallurgical work, *Anweisung zur Metallurgie* and the famous *Specimen Beccharianum*, a long preface/commentary on Becher's *opus magnum*, the *Physica subterranea*.[22] Stahl's status and influence became so essential to these publications that after 1718 Eyssel consistently started the titles of his Stahl series with the author's official title.

Thus by 1720, Eyssel had established a successful business of publishing Stahl's academic works in German, and the experience of the *Observationes clinicae* that had already received two editions (and eventually went through two more by 1735), was suggestive enough to Eyssel that another hitherto unpublished textbook by Stahl, namely the *Chymia rationalis*, would also be a commercial success. That notion turned out to be well founded, for the *Chymia rationalis* saw its first reprint in 1729, and was reprinted again by a Schönemarck, also a Leipzig publisher, in 1746. Altogether Eyssel published more than 15 titles of Stahl's works in about 25 editions. His effort constituted a serious campaign to introduce Stahl's works, the majority of which had been published in Latin due to their academic nature, to a German-reading market.

1720: *VERBESSERUNG DER METALLEN* AND ITS PUBLISHER

Stahl's other 1720 alchemical publication was the *Verbesserung der Metallen*. This work actually first appeared in print in 1703 in Latin as an article in the journal *Observationum selectarum ad rem litterariam spectantium*, a journal often identified by the place of its publication as *Observationes Hallensis*. The journal was co-edited by Stahl and his colleagues Christian Thomasius (then already a famous jurist and philosopher in Germany) and Johann Franz Buddeus, who later rose to prominence as a Lutheran theologian. The article Stahl contributed as "De metallorum emendatione" was republished in German by Roth-Scholtz as *Verbesserung der Metallen* in the form of an independent octavo.

The *Verbesserung der Metallen*, or "De metallorum emendatione," sheds important light on Stahl's thinking about alchemy and alchemists at roughly the middle of his career in Halle. He began with a consideration of the origin of alchemy and its beginnings in Europe, lamented the craze for chrysopoeia in Europe, and yet accorded this art a due place. He classified pursuers of chrysopoeia, and suggested the qualifications necessary to be a good seeker of transmutation. Though maintaining a cautious tone, Stahl was open to accepting the truth of this art.

Stahl first considered the country of origin for the art of perfecting metals. While in his Jena lectures he taught that alchemy was of Egyptian and Arabic origin,[23] here he suggests that it originated instead in China. He points out the great inventions of the Chinese: printing, gunpowder, and the manufacture of paper. Along with these great inventions, he says, the Chinese have practiced since very ancient times a double art that deals with the transmutation of baser metals into perfect ones on the one hand and with the prolongation and rejuvenation of life by medical or pharmaceutical means on the other.[24] This alchemical art was transmitted to Europe by the Arabs, Persians, Parthians, and Indians, who were active in long and extensive commerce in the areas between Europe and China.[25] In Europe, the investigation of these things cannot be earlier than the circulation of the works under the names of Isaac Hollandus, Basil Valentine, and Raymond Lull, and certainly not, as some claim, centuries before Paracelsus's time.[26]

While accepting the reported successes of various alchemical authors, Stahl also enumerates the vices of the craze for chrysopoeia. Among those whom Stahl identifies as having documented chrysopoetic success are Johann Zwölffer (or Zwelfer, 1618–68), Johann Joachim Becher, Daniel Georg Morhof (1639–91), and Baron Wilhelm von Schröder, all seventeenth-century figures.[27] The desire for gold, however, has given rise to impostures, frauds, apprehensive labor, wasted time, excessive expenses, destroyed health, neglect of family, and ruined reputation. Although the Roman Catholic Church banned it as heretical under pain of denunciation or excommunication, this lamentable condition grows instead of diminishing in the Christian world.[28]

Stahl distinguishes between two kinds of pursuers of chrysopoeia, whom he calls the universalists and the particularists. The universalist wants nothing but the entire and absolute transmutation of one thing into another. The particularists, who are content with a smaller profit, can further be divided into two different groups. One consists of those who are happy with a considerably smaller share of the gain that the universalists seek to make. Not much more respectable than the universalists, they are likewise destined to be deceived by their hopes and therefore to be ruined. The group of particularists that wins Stahl's approval, however, is characterized by their honest attitude and modest management style, so to speak. Compared to the greedy alchemists, they are endowed with an evenness of the mind, thus carefully offsetting their expenses with some honest profit and interest. They can only be wealthy men who have plenty of "capital" or assets.[29] There Stahl implies that the financial status is a precondition for the sound mental and moral quality of the chrysopoetic pursuers.

For guides to chrysopoeia, Stahl names Fallopius, Kessler, Glauber, and especially Becher, a personality famed as a powerful figure at European courts and an influential chymical author of German origin.[30] He cites Becher's *Chymischer Glückshafen*, *Physica subterranea*, and *Minera arenaria*, where Becher promised that a huge fortune can be gained by an easily obtainable and "infallible" method. Concerning the credibility of Becher's assertion on the profit of his method, Stahl writes:

> I do not fear to say that, as I have not hesitated to save the honor and credit of this author in other things, so I do not hesitate at all to subscribe to his assertion on this condition that, although I am neither confidently informed of his real method, nor can I consequently show what he is actually able to achieve, yet I testify without fear, no less on his promise than on his method as he has stated it, that at least in general it is not completely without hope of usefulness,

and is able to do something, indeed so much that it may finally be worth the effort, even if it does not correspond to the exact calculation of the author.[31]

Though approving Becher's alchemical knowledge, Stahl in effect disagrees with him on who should pursue the work. Becher spent his life approaching ruling princes at European courts, trying to secure their support for chrysopoetic projects.[32] In contrast, Stahl maintains that "private persons," as he calls them, are better than princes or magnates who have responsibilities in state administration. Private persons can try this art as honestly as they try any other art. In contrast, attempts by great lords are not recommended because the outcome may often be unfortunate and changeable. Moreover, the great work requires a great deal of operators, apparatus, and labor, thus not befitting their status.[33] Yet although it is not appropriate for men of high offices to try this, it is not wise to ban it with state power. While people who use hasty judgement may make such a suggestion, no ruler should prohibit his subjects from trying their hand at transmutation, especially when he is unable to prevent other countries from discovering the true knowledge of this lucrative art. Thus international competition justifies citizens' alchemical pursuits.[34]

Stahl goes on to elaborate the qualifications of those suited to this job with an argument on the return rate of the alchemical investment. He calculates that the profit from chrysopoetic work would never be the tens of thousands of times the investment as its promoters boast. But if a private person is willing to invest what Stahl calls an idle capital of a thousand Thalers, and entrusts others to do the work, he would be able to collect between 150 and 200 Thalers a year after paying all the expenses, and would earn somewhat more if he does it himself. But either way, the money has to be his own so that he can acquire the necessary capital honestly without doing damage to anyone else's financial security. As magnates or state rulers presumably always have difficult cash flows, it is wise for them to leave it to wealthy private men.

The time, 1703, of the first publication of the Latin original of this text – "De metallorum emendatione" – is significant in at least three regards. First, Stahl's favorable consideration of China as the country of origin for alchemy reflects the interest of his age. Leibniz had just published his correspondence with Jesuit missionaries about China as *Novissima sinica* (*The Latest News from China*) in 1699, and Stahl's colleague Thomasius also published his own favorable view of China in 1689.[35] Second, the article came out in the same year that Stahl republished Becher's *Physica subterranea*, to which he appended his commentary *Specimen Beccherianum*. Both the "De metallorum" and the *Specimen* show Stahl's admiration for Becher. While in the *Specimen* Stahl demonstrated his grasp of Becher's subterranean physics, his respectful reference to Becher in "De metallorum" served in no small sense to promote Stahl's contemporaneous republication of the *Physica subterranea*, and showed the great degree to which he identified himself as Becher's close follower. Third, just before Stahl published "De metallorum emendatione," the community of the University of Halle was engaged in a high-profile public discussion on the alleged success of the young alchemist Johann Friedrich Böttger (1682–1719) in transmuting an imperfect metal into gold. Böttger's story received tremendous attention from the powerful and the learned. Two of the most powerful states of Protestant Germany, Brandenburg-Prussia and Saxony, competed to capture this reported adept. The community of Stahl's university seems to have been eager to discuss Böttger's case, as it became the subject of an academic disputation open to the entire university community. This disputation was organized and chaired by Buddeus, a co-editor of the above-mentioned journal *Observationes Hallensis*, and who, even more importantly, was then serving as the Prorector, the highest office of the university.[36] The title of the thesis was "Whether alchemists should be tolerated by the republic?" and the thesis stated in essence that, although the frauds and deceptions of malicious alchemists should be prevented with severe punishments, the polity should tolerate its practice, for no human knowledge had disproved the possibility of chrysopoeia.[37] Stahl, in his 1703 article, thus agrees with Buddeus that the chrysopoetic art should be allowed, while he qualified his support with stipulations about the social and economic status of legitimate pursuers.

The 1720 publication of the German translation *Verbesserung*, however, places this text in a new context, one that was tied to its publisher's project of alchemical publications. Roth-Scholtz had been undertaking his project with the publishing house in Nuremburg that he took over as an heir to his father-in-law Johann Daniel Tauber.[38] His *Bibliotheca Chymica*, an ambitious catalogue which aimed to list all chymical books ever published, was well underway. Indeed, he boasted in a footnote to the preface of *Verbesserung* that he had compiled more than 5000 works for the catalogue that he would publish soon afterwards.[39] The preparation of the catalogue went hand in hand with his publication of works by famous alchemical authors. Before he published Stahl's *Verbesserung*, he had published in the late 1710s works of the best-known alchemical authors, including Basil Valentine, Michael Sendivogius, and Becher, all in German.[40] Once Stahl stood out as one of the most powerful chemico-medical figures in Protestant Germany after his establishment at Prussian court, he became an ideal person to be added to Roth-Scholtz's publication list.

Enlarging on Stahl's title at court and his distinguished status in medical and chemical studies, Roth-Scholtz emphasized the value of his translation. This text has been published three times, the publisher indicates, but always in Latin.[41] Yet this thoughtful (*sinnreich*) work of so prominent an author deserves to be read by the learned and unlearned (*Gelehrten und Ungelehrten*) alike. He notes that Stahl has hinted at his knowledge of chrysopoeia here and there in his writings, although he has concealed his secrets in such a way that he wants to prevent the reach of those who seek it "for bread," and instead to reveal them only to honest pursuers of truth.[42] Roth-Scholtz nevertheless leaves his readers with no doubt that Stahl, now a prominent chemico-medical figure, is a steadfast patron of alchemy. The publication of *Verbesserung* provides not only a proof of Stahl's positive view of alchemy but also moral support for Roth-Scholtz's own alchemical publication project.

Thus Roth-Scholtz appropriated *Verbesserung der Metallen*, which recorded Stahl's thinking on alchemy in 1703, to his own favor. At the time of its first publication, Stahl viewed China as the country of origin for alchemy, suggested "private persons" as the ideal pursuers of chrysopoeia, presented an interesting consideration on the "return rate" of alchemical investments, supported chrysopoetic pursuits on the grounds of international competition, and identified alchemical authors he respected, among whom Becher stood highest. Some of these ideas reflected the intellectual, personal, or local context at the time of its first publication. Roth-Scholtz, however, included this text in his publication project to support his own efforts.

STAHL'S ALCHEMICAL THINKING IN TRANSITION

The two above-mentioned publications of 1720, both favorable to alchemy, presented to German readers an apparent anachronism. It is all too often taken for granted that one's textual production is truthful to one's thinking at the time of publication; at least it is rarely expected that a publication contradicts its author's thinking. Enthusiastically promoting their publications of Stahl's alchemical works, both Eyssel and Roth-Scholtz were silent about the fact that Stahl's reservations about chrysopoeia had grown greater in the late 1710s and then became outspoken criticisms by the early 1720s.

Beginning in 1718, a wave of Stahl's newest chymical works was entrusted to the press. Unlike the translations that Eyssel and Roth-Scholtz published, these works were Stahl's most recent writings and were published directly in German. The first of them was Stahl's well-known treatise on sulfur.[43] In this work he was ambivalent about chrysopoeia. On the one hand he seriously considered the truth of the transmutation of an imperfect metal into gold by a tincture, that is,

> the substance that gives gold its color, serves at the same time to bind more intimately the other principles, to render them finer, to render them suitable to occupy a smaller space, above all since silver, when it has been transmuted into gold by art, ought to surpass the same mass smaller in the previous weight [i.e., keeping the same weight in a smaller volume].[44]

Yet he also gave serious warnings that Kopp interprets as "considerably less comforting and encouraging for the alchemists (*für die Alchemisten erheblich weniger tröstlich und aufmunternd*)."[45] In his treatise on salts that was published five years later, Stahl became even more outspoken with his warnings. He accepted the transmutation or the Verbesserung of metals as an art (Kunst), and yet seriously doubts its actual plausibility:

> I can affirm ... from manifold examples that have come to my experience, that I have the most honest reasons in the world to believe that such great credulity concerning the affirmation of this question [of truth of chrysopoeia] is a very ungrounded, and, as so many hundred examples reveal, highly offensive and harmful matter."[46]

A similar judgment was repeated in a long comment that Stahl was invited to write on a metallurgical work by Becher. The third part of *Billig Bedencken, Erinnerung und Erläuterung über D. J. Bechers Natur-Kündigung der Metallen*, also published in 1723, was dedicated to Becher's discussions of metallic transmutation. Although in the first part Stahl reaffirmed his praise for Becher's theory of chemical principles, in the third part he withdrew his trust from Becher's alchemical teachings. He asserted that he was not convinced by Becher's propositions of the supposed two powers of the *Tinctur*, that is, its ability to transmute metals and its near-miraculous medical power. He related that "it is namely the most doubtful question: whether it is, or has ever been, true that it is possible to find or produce an essence (*Wesen*) of the kind and power as the *Tinctur* is described to be?"[47] Thus for Stahl in 1723, the truth of chrysopoeia was very doubtful.

The three chymical works examined in this section indicate another new trend in Stahl's publication history, namely, publishing directly in German. The Stahlian publications by Eyssel and Roth-Scholtz had already set off one side of this trend, namely, introducing academic works, previously written or taught in Latin, to the German-reading market in translation. In contrast, these three new texts were published directly in German. Their publishers were not presses specializing in popular works, and the substance and discursive style of these works are not much different from those of Stahl's previous works written in Latin at his academic post (such as *Specimen Beccherianum*). Thus they cannot be claimed to be distinctly popular works.[48] Their appearance nonetheless signals that the publishers and the author thought it worthwhile – economically or intellectually – to present these texts directly in German, thus appealing to a readership larger than the exclusively Latin readership. The continued success of Eyssel's Stahl translation project seems to suggest that there existed a market for chemico-medical publications among those who were eager to read German academic works previously available only to Latin readers. The three 1718 and 1723 publications indicate that the Latin readership could be skipped, presumably because more and more Latin readers were willing, and certainly able, to read the German publications of important learned authors without feeling disgraced. The two sides of this new trend appear to have developed hand in hand, beginning in the mid-1710s.

The publications of this period happen to have recorded the great shift of Stahl's alchemical thinking during 1718–23, and thus created the apparent anachronism. During this period, Stahl made an about-face to become a disbeliever of chrysopoeia from being a believer. Although he left behind almost nothing that documents the personal or social factors that occasioned this shift, his writings from this period make it clear that the early works supportive of chrysopoeia, republished by Eyssel and Roth-Scholtz in 1720, had by then ceased to be accurate reflections of his thought. While there is no evidence that the two publishers were aware of this shift, it would soon become clear that their publications contradicted Stahl's current thinking.

"BEDENCKEN VON DER GOLD-MACHEREY": OPEN CRITICISM OF ALCHEMY AND THE RESHAPING OF THE BECHER-STAHL RELATIONSHIP

Stahl's best-known criticism of alchemy, "Bedencken von der Gold-Macherey," was a commentary that served as the foreword to the 1726 republication of Becher's collection of chymical processes, *Chymischer Glücks-Hafen, oder Grosse Chymische Concordantz*.[49] This was certainly not the first time that Stahl published a commentary as an appendix to a reprint of Becher's work. But this time Stahl seemed ready to make a clear statement on his relationship both to alchemy and to Becher. Here he lashed out against alchemy, blaming it for the disruption of civil order, negating its moral value, and rejecting all the theoretical and practical possibilities of transmutation that had been previously proposed.

Stahl began by differentiating between *Alchymie* and *Chymie*, a distinction that, as Principe and Newman point out, was quite new at the time.[50] For Stahl, alchemy is a work based on mere experience and long-lasting exercise (*Übung*), and is a confused, obscure, and futile undertaking. It is characterized by *Goldmacherei*, namely, chrysopoeia. *Chymie*, on the other hand, is a rational, well thought out (*wohlbedachtliche*), and intellectual (*wohlverstandene*) study and work that leads to thorough (*gründliche*) knowledge.[51] The objective of the "Bedencken" is to examine the fallacies of *Goldmacherei* in four aspects: first, in regard to civil order and *Polizei*; second, its moral merit; third, the theoretical and practical possibility of transmutation; and finally, the source of alchemical instructions.

Chrysopoetic pursuits, according to Stahl, become a concern of the *Polizei* because of the social and economic consequences of the unrealized hope for chrysopoeia. It drains too much from the resources of individual persons and the polity and helps produce deception and fraud, thus endangering the civil and economic stability of society. It is this concern with civil and economic order that relates alchemy to *Polizei*, a key notion of cameralism. For the cameralists, *Polizei* referred to the maintenance of social and especially economic order of a polity by using any policy measures available to the state, whether political, financial, police, or even military.[52] Stahl's inclusion of *Polizei* in his discourse appears to indicate that the value of the cameral sciences were on the rise not only in the German intellectual world in general, but also in his own mind in particular.

The moral value of *Goldmacherei* is as doubtful as the physical value of the precious metals is unjustifiable. For Stahl, gold and silver are not necessary to sustain the human body or life. Since their practical use is not as great as that of iron, copper, or zinc, their high market value does not derive from their use for life, but instead depends on their rarity. Consequently, this value would drop if their availability increased. Indeed, says Stahl, the value of silver (and its purchasing power) have actually dropped to only one-tenth of what they had previously been due to the large number of silver mines discovered in preceding centuries. Hence, the value of the precious metals is not even durable. Only a true understanding of their actual value reveals the immorality of the craze for them that gives rise to wasteful pomp and bad management of personal wealth.[53] Indeed, the moral status of chrysopoeia, to which the craze for precious metals is tied, cannot be justified by the alleged contribution that the apologists of alchemy often use to justify its worth; they argue that alchemical attempts have led to the discovery of very useful pharmaceuticals (*Arzney-Mittel*). But although these pharmaceuticals are indeed useful, this value is not to be credited to alchemy, but rather to respectable chemical handiwork (*erbaren Chymischen Hand-Arbeiten*).[54] The alchemists "through this art of futile undertakings never bring about or produce anything true."[55]

By this point Stahl has come to believe that under no circumstances can one species of metal be produced out of another, and he refutes transmutational propositions with reflections on empirical reports or observations. Different metallic species can replace, transpose, and mix with one another in chemical processes; they can be altered so much into raw, sharp, and fluid crudeness that they appear destroyed. But these observations constitute no proof for transmutation between species. For Stahl, belief in transmutation seems to rest on the assumption that metallic ores ennoble themselves in subterranean veins. Yet the claim for the ennoblement of metallic veins could mean little more than that ores found in some locations are not as good in quality and quantity as those of more productive veins.[56] A derivative, but more specific, assumption applies to mines in snow-capped mountains. It is reported by master smelters that in those mountain mines a vertical stratification is formed; on top are rich iron mines, in the middle is iron mixed with silver, and the deeper one goes, the richer the silver mines become. This observation is used to prove that the iron ore improves itself into silver. But relying on his metallurgical knowledge, Stahl instead proposes that the phenomenon is caused by the presence of cobalt, which is known to exist in greater abundance deeper into the earth. Cobalt is, as Stahl calls it, a deeply polluted silver essence (*ein tieff-verunreinigtes Silber-Wesen*), suggesting its frequent natural association with silver – an observation proved accurate by today's scientific knowledge. Silver is made "fugitive" (*flüchtig*) by forming a compound with cobalt and this compound or alloy is eaten away by salts, especially arsenic salts. As the solutions of these compounds drain downward, sediments eventually form in the deeper parts of the mountain veins.[57] The stratification of different metallic ores is thus the result of erosion and sedimentation, not of natural transmutation.

Stahl further cites metallurgy to argue that no transformation (*Umsetzung*) of one metal into another can be seen to occur. Although copper and iron ores are found to be mingled so closely that the naked eye cannot distinguish one from the other, their mingling is not an inner integration that transforms species. Therefore, smelters can retrieve one or the other with the acids often used in ore refining. Thus chemical substitutions, in our modern concept, are well recognized by Stahl, but they are not to be mistaken as metallic transmutations.[58]

Then Stahl discusses the coloring essence (*farbhafftiges Wesen*), that is, tincture. The coloring essence is thought to be contained in iron and copper, and able to furnish not only the yellowish color to other metals, especially silver, but also to complete the essence of gold. Yet Stahl maintains that even though human beings have found, melted, and processed an immense amount of iron ore, especially the iron-rich ores in red earths (probably oxidized iron in our terms) as well as a comparably great amount of copper, experience has yielded no successful tincturing. Likewise he considers empirically ungrounded the assumption that by making essential metal parts subtle another metallic species of different combination (*Vermischung*) can arise. This new weight proportion must be produced by an "effusive force" (*überschwengliche Krafft*) that has never been observed.[59] Stahl then considers whether there might be an artificial way to achieve transmutation even if there are no natural examples of its occurrence. Although such possibility cannot be ruled out, credible examples are not yet available. For instance, stories of successful transmutations reported by Johann Christian Orschall and Johann Kunckel simply give too little credible detail to win trust.[60]

The last topic for Stahl's criticism is where and from whom one could learn such an art. The short answer is nowhere and from no one. Although Olaus Borrichius endeavored to make an ancient origin for alchemy possible, none of his supposed ancient sources are reliable. Indeed, many of the figures in those sources may never have existed. Characteristic of such sources is the alchemical text *Turba philosophorum*, whose characters are incredible figures such as Arab kings, Romans, Aristotle, the Jewish prophet Moria, Moses' sister, her student Aros, and even Moravians. Recent texts ascribed to Lull, Villanova, Albert Magnus, and Flamel are no better sources and have no more credible evidence. Basil Valentine and Isaac Hollandus are persons who actually lived, and yet like the others, they brought to light no distinctly trustworthy knowledge of transmutation.[61]

Finally, Stahl turns to general problems shared by alchemical authors. First, even though they claim to imitate nature, their intention is not to understand nature's way of generating metals. For example, they never note how little gold or even silver there is in nature compared to other metals. It is simply their false conception of natural working that the rays of the seven planets would give rise to metals deep down in the earth.[62] Second, likewise they have a wrong belief that a humid, fatty, and smutty mist (*Dunst*) in the veins of mines produces new strokes of ores, and even sprinkles ores with gold. For Stahl, this humidity is rather a vitriolic essence (*Wesen*). Far from being the "protoplasm," so to speak, that gives birth to metals, this vitriolic mist is in fact not found with all metallic ores. It appears only with iron- and copper-rich ores, and produces vitriol and alum instead of gold.[63] Similarly the belief that cinnabar ores produce gold is never supported by credible reports.[64] The alchemical authors' strongest argument has been that the transformation of metals has actually occurred, yet such occurrences have not been confirmed anywhere with sufficiently authenticated evidence.

Regarding his motivation to write this preface, Stahl asserts that

> I have upon the republication of Becher's *Chymische Concordanz* [i.e., *Chymischer Glücks-hafen*], at the request of the publisher, wanted to bring forth my present thinking especially with the intent that people might at least care to cultivate more rational thinking and reflection, in order not to engage in hopes so little grounded and confirmed .... but instead apply their studies and exercises, leisure time, and spare funds in accordance with nature, reason, true art, and dexterity. Thereby some rational delightfulness, clear science, reasonable advantage, and use can ensue through true chemistry and its well-used application.[65]

One may well imagine that this publisher, the little known Halle-based Ernst Gottlieb Krug, wanted Stahl to say something good about the text. His humble wish obviously was not fully granted by the Royal Physician, who may have had some influence on privileges for publication in Halle where the press was based. Stahl adds however that Becher's collection of chymical processes is not useless. It provides readers with an opportunity to become proficient in handiwork such as mining, tarring, and brewing, and to gain deeper insight on the internal working of metallic and mineral bodies.[66]

For readers who closely followed his writings, Stahl feels it necessary to explain the transformation he has undergone. He emphasizes the strong warning against chrysopoeia of his recent works, namely, in his treatises on sulfur and on salts and in his commentary on Becher's *Natur-kündigung der Metallen*. He denies the allegation (alluded to by Roth-Scholtz) that he has on occasion noted the probability of transmutation. Stahl argues that in almost all places where he wrote positively, he rightly added the condition that if their boasts should indeed come true, they most probably would be grasped one way or another. Besides, he approved only specific experiments, not the entire work or theory of such authors.[67] Only in his last paragraph does Stahl concede that metallic transmutation cannot be regarded as absolutely impossible, although he avers that he has no knowledge of gold-making and has never had the tincture. Even if chrysopoeia might not be impossible, it would not be realized until vastly more well-experienced work, patience, time, exact observation of the material circumstances, and repeated tests have been applied. In closing he cites Kunckel's warning that inexperienced and impatient people must not take up such work.[68] Thus concludes Stahl's most direct, and most explicitly critical, text on alchemy.

The publication of "Bedencken von der Gold-Macherey" redefined Stahl's relationship to alchemy and repositioned him in the history of chymistry. Stahl makes no reference to the textbook for his course given more than 40 years earlier. His outspoken criticism of Isaac Hollandus in the *Bedencken* in effect disavowed his former ties to authors like him, ties that were reinforced by appending Hollandus' alchemical text to the 1720 *Chymia rationalis*. Likewise, Stahl's 1726 criticism of chrysopoetic theories and practices revoked the other appendix to the earlier book, his treatise on mercuries and the philosophers' stone. In 1726, Stahl makes no reference to the 1703 article in *Observationes Hallensis*, where he lent support to chrysopoeia, nor to Roth-Scholtz's translation of it, the *Verbesserung*. Instead of supporting or apologizing for his previous views, he just quietly disowns them. That was certainly no supportive message to the publisher Roth-Scholtz, who expanded his alchemical publication project by co-opting Stahl as an intellectual patron. But the 1726 "Gold-Macherey" also marks Stahl's departure from his previously self-proclaimed role as a follower of Becher. He had previously promoted his own status in chemistry by associating himself with the famous Becher. The best known of these efforts was the 1703 republication of Becher's *Physica subterranea*, for which his *Specimen* served as the manifesto of a follower. Indeed, seen through the eyes of Stahl's students, or even of twentieth-century historians, this relationship was so close that the chemistry of Becher and Stahl have since formed a uniform "school."[69] His association with Becher, of course, did not escape the attention of publishers. His *Specimen Beccherianum* was published in German translation by (once again) Eyssel in 1720, and in 1723 he was asked to comment on Becher's *Natur-Kundigung der Metallen*. Finally, when a Halle publisher wanted to republish Becher's collection of chemical processes, it was again Stahl that he approached for a foreword, although one can reasonably doubt whether the critical foreword that Stahl actually provided worked in favor of the book's sales. Stahl's 1726 diatribe against chrysopoeia, "Bedencken von der Gold-Macherey", however, is the clearest declaration that he no longer subscribes to Becher's alchemical ideas, even though he had been willing to accept his predecessor's theory of chemical principles and practical chemical knowledge. Thus Stahl parted in public with a master to whom he once made efforts to associate himself. This rupture occurred at a time when Stahl had accomplished at least as much as a chymical author and at court as Becher had done, and thus no longer needed to depend on Becher's authority in the chymical world.

CONCLUSION: TEXTS, AUTHORS, AND PUBLISHERS OF ALCHEMY IN THE CONTEXT OF EARLY EIGHTEENTH-CENTURY GERMANY

This paper has reviewed three texts by Stahl that deal directly with chrysopoeia. One represents his earliest teaching from 1684, the second his position in the middle of his tenure at Halle, and the last his views close to the end of his life. It is obvious that Stahl turned from a believer to a disbeliever; this observation alone is significant for our historiography of Stahl. Although Kopp indicated this shift as early as the nineteenth century, neither he nor other historians have taken note of the apparent anachronism presented by Stahlian publications; namely, that the two pro-chrysopoeia publications of 1720 are out-of-date – as earlier compositions they no longer represented Stahl's actual thinking at the time of their publication. Their publication date of 1720 thus gives the unwary historian a false reading of Stahl's opinions at that time. Nobody, including this author, has yet found source material that explains why Stahl changed his mind about alchemy. This paper, however, reconstructs the context that explains the origin of this apparent anachronism, that is, the fact that Stahl's earlier, more positive views of alchemy were recycled and published outside of his control and at a time when he had become a critic. This later publication of two early alchemical works by Stahl involves two publishers with different agendas in an age when the introduction of important academic authors' works to the German-reading public was becoming a trend.

Stahl's relationship with these two publishers, the Leipzig-based Eyssel and the Nuremberg-based Roth-Scholtz, had different endings. Both publishers were entrepreneurial to a considerable degree, and both wanted in 1720 to bring Stahl's earlier alchemical works to press, though in so doing they created the anachronism in regard to Stahl's position on chrysopoeia. What happened to these publishers after they published teachings that Stahl no longer embraced? Eyssel continued to translate and to publish Stahl's academic works well into the 1730s; many of these were Stahl's dissertations. For a collection of gynecological dissertations published in German in 1724, Stahl actually gave Eyssel a new text that served as a postscript to the collection and provided a review of the development of his medical thinking.[70] That Eyssel received this new text suggests that his relationship with Stahl continued on mutually acceptable terms after the publication of *Chymia rationalis* in 1720. It is easy to see that a continuing and large-scale introduction of Stahl's academic works to the German-reading public did not violate Stahl's intellectual and economic interests.

Roth-Scholtz's relationship with Stahl, however, was not as successful. He managed to publish only one more text by Stahl, a short text on chemical pharmacy.[71] He included no work by Stahl in his *Deutsches Theatrum Chemicum*, an anthology of great alchemical works in German, even though he produced a German translation of the above-mentioned thesis chaired by Buddeus presented to the University of

Halle in the wake of Böttger's reputed success of transmutation – in fact, Roth-Scholtz placed it as the opening piece of the *Theatrum*. We have no direct evidence to indicate why Stahl's relationship with Roth-Scholtz did not continue. Yet it is obvious that once Stahl became a convinced disbeliever of alchemy, it was not in his interest to support Roth-Scholtz's agenda of alchemical publication.

The context of Stahl's chymical publications shows how texts can be appropriated by publishers, students, other authors, and so forth, once they have gained their own existence. Publishers printed texts to advance their own economic or cultural projects, either with or without their authors' cooperation. Authors used texts in a variety of ways as well, for besides the ordinary functions of publication – teaching, addressing, arguing, defending, and so forth – authors also reprinted, commented, or prefaced texts by other writers with whom they wanted to associate or dissociate themselves. This is what Stahl did with Becher's works. On the other hand, the writings put into print by Stahl's students, such as Ulau and Carl, who owned or found compilations of notes from their former professor's lectures, functioned either simply to publicize the master's teaching or to contribute to their own personal advantage; thus it is not only printed texts that can be appropriated. What makes Stahl's case both so complicated and so interesting is that manuscripts from his lecture courses (that were not exactly his own composition) circulated outside of his reach. The underlying condition for the complexity of Stahlian publications is that authors do not always have complete control over the publication or republication of their works, especially not in early eighteenth-century Germany where no central authority could effectively enforce anything close to a modern copyright over the hundreds of states within the Holy Roman Empire.

In the early eighteenth century, the publication of academic works in German was a recent phenomenon and Stahl was a chymical author at the forefront of this new trend. Of course, the publication of chymical works in German was not new. Becher, and especially his older contemporary Glauber, published many works in German. Yet, although Becher had an image as a "learned" figure, neither he nor Glauber wrote their works as academics or for an academic audience. Academic chemico-medical works were never written first in German so long as Latin remained the predominant academic language in Germany. Even German translations of such works were scarce throughout the seventeenth century. Recall that even Libavius's famous *Alchemia* was not translated into German until the twentieth century.[72] Daniel Sennert, Werner Rolfinck, and Georg Wolfgang Wedel, who represent three generations of the most important academic chemico-medical authors of seventeenth-century Germany, saw no German translations of their works before 1700. The trend towards translation seems to correspond with the introduction of German into academic life, started first by Christian Thomasius's revolutionary move to give university lectures in German in 1687, and to publish works of academic nature in German. Soon to become one of the best-known academic authors in Germany, Thomasius still wrote a great deal of his academic work in Latin, even though a considerable number of them became available in German shortly after their publication from the 1690s on. Stahl's works were translated later in comparison to Thomasius's, but his German publications were about as early as those of Christian Wolff, probably the most important German author of the first half of the eighteenth century. Wolff, 20 years younger than Stahl, nonetheless led the trend for publishing textbooks directly in German. His work known as *Deutsche Logik*, for example, was first written in German and published in that language in 1713.[73]

One might ask, if Stahl turned from a believer of alchemy to a critic, how did he view the publication of his older works that contradicted his current thinking? The author of the annotated bibliography of Stahl, Johann Christoph Goetze, suggested that the 1720 and 1723 publications of Stahl's Jena textbook received some sort of permission from the Royal Physician.[74] Yet shortly before his death, Stahl wrote a letter to one of his closest disciples, Johann Juncker, then professor of medicine at Halle, asking him to testify to the "alchemical calamity" (*alchemicam calamitatem*) in the second volume of *Conspectus chemiae theoretico-practicae* that Juncker was preparing. Following his former teacher's request, Juncker not only castigated alchemy but also reproduced part of Stahl's letter (written in German) in the preface to the second volume. In the letter, referring to his *Collegium Chemicum* of 1684 "that Carl afterwards edited," Stahl confessed that, "then at the age of twenty-five I was not so perfectly free from all credulity of that sort." In a defensive tone, he said that many points in the textbook "may not be completely in vain or false," and yet he noted this did not "apply to the foolish transcendental hope or imagination of gold-making."[75] Careful not to identify his earlier teaching as literally foolish, Stahl at least conceded that as a youth he was contaminated by the credulity that characterized believers in chrysopoeia; but this self-criticism was not made public until 1738, four years after his death.

ACKNOWLEDGEMENT

The research for this paper was supported by Taiwan's National Science Council, grant nos. 93–2411-H-001–062 and 94–2411-H-001–063.

NOTES

1 See, for example, Jon Eklund, "Architects of the Mature Phlogiston Theory," chapter 2 of his "Chemical Analysis and the Phlogiston Theory, 1738–72: Prelude to Revolution" (Ph.D. diss., Yale University, 1971), and James R. Partington and Douglas McKie, "Historical Studies on the Phlogiston Theory," *Annals of Science* 2, 1937, 361–404; 3, 1938, 1–58, and 4, 1939, 113–49.

2 Bernadette Bensaude-Vincent, *A History of Chemistry*, trans. Deborah van Dam (Cambridge, MA: Harvard University Press, 1996), 60–61.

3 For example, Allen G. Debus, *Chemistry and Medical Debate: Van Helmont to Boerhaave* (Canton, MA: Science History Publications, 2001); compare with Karl Hufbauer, *The Formation of the German Chemical Community (1720–1795)* (Berkeley, CA: University of California Press, 1982), 8–11, 167–68, and Johanna Geyer-Kordesch, "Chemie und Alchemie: J. J. Becher, G. E. Stahl, J. S. Carl, und J. C. Dippel," in *Johann Joachim Becher (1635–1682)*, ed. Gotthardt Frühsorge and Gerhard F. Strasser (Wiesbaden: Otto Harrassowitz, 1993), 127–42.

4 James R. Partington, *A History of Chemistry*, 4 vols. (London: Macmillan, 1961), 2:686.

5 Hermann Kopp, *Die Alchemie bis zum letzten Viertel des 18. Jahrhundert*, 2 vols. (Heidelberg: C. Winter's, 1886), 1:69–75.

6 William R. Newman and Lawrence M. Principe, "Alchemy vs. Chemistry: The Etymological Origins of a Historiographic Mistake," *Early Science and Medicine* 3, 1998, 32–65.

[7] *Chymia rationalis et experimentalis; oder, Gründliche, der Natur und Vernunfft gemässe und mit Experimenten erwiesene Einleitung zur Chymie* (Leipzig: Caspar Jacob Eyssel, 1720). I cite the third edition of 1746, done at Leipzig by Wolfgang Heinrich Schönermarck.

[8] *D. Georg Ernst Stahls … Gedancken von Verbesserung der Metallen, und wie man einen mässigen Gewinnst davon ziehen könne* (Nürnberg & Altdorff: Johann Daniel Taubers sel. Erben, 1720).

[9] Georg Ernst Stahl, "Bedencken von der Gold-Macherey, an statt einer Allgemeinen Vorrede und Einleitung, über verschiedene, weitläufftig dahin an- oder verleitende Discurse und Processe," in Johann Joachim Becher, *Chymischer Glücks-Hafen, oder Grosse Chymische Concordantz und Collection* (Halle: Ernst Gottlieb Krug, 1726). Since the text is unpaginated I will use a pagination that counts the title page as 1.

[10] By "chymical" matters I, following Newman and Principe's proposal, mean things whose nature runs across the divide between the now stereotyped view of chemistry (rational and scientific) and that of alchemy (mystical and pseudo-scientific) at the early modern age when alchemy and chemistry were essentially synonymous with each other. See Newman and Principe, "Alchemy vs. Chemistry."

[11] Hermann Kopp and K. C. Schmieder suggest that this text was first published as a dissertation in 1682, a time when Stahl was still working on his M.D. I doubt the truth of their suggestion. First of all, Friedrich Roth-Scholtz seems to have had no knowledge of such a dissertation when he enumerated previous editions of the title in his foreword to *Verbesserung*. Second, I have not been able to find such an edition in any library catalogue. Third, it is not mentioned in the bibliography of Stahl prepared by his contemporary Johann Christoph Goetze, *Scripta D. Georg. Ern. Stahlii*, 2nd ed. (Nuremberg, 1729). See Kopp, *Alchemie*, 71, and Schmieder, *Die Geschichte der Alchemie* (Halle: Waisenhauses, 1832), 508–09.

[12] Stahl was appointed in 1715 Chair of the Berlin Medical Board. This board was reorganized in 1725 to be the Higher Medical Board for all of Brandenburg-Prussia. He was also a member of the new Sanitation Board in 1719–34. See Hufbauer, *Formation of the German Chemical Community*, 167.

[13] Peter Shaw, ed. *Philosophical Principles of Universal Chemistry, or, The Foundation of a Scientifical Manner of Inquiring … Drawn from the Collegium Jenese of Dr. George Ernest Stahl* (London: J. Osborn and T. Longman, 1730).

[14] *Georg. Ernest. Stahlii Fundamenta chymiae dogmaticae & experimentalis … annexus est ad coronidis confirmationem tractatus Isaaci Hollandi de salibus & oleis metallorum* (Nuremberg: sumptibus Wolfgangi Mauritii Endteri haered., typis Johannis Ernesti Adelbulneri, 1723).

[15] Stahl, *Chymia rationalis*, translator's foreword, fol. 5r.

[16] Stahl, *Chymia rationalis*, 459–520.

[17] Ibid, "Vorrede," unpaginated.

[18] Stahl, *Observationes clinicae, a viro experientissimo privatis praelectionibus quondam traditae, … zum Druck befördert von D. Gottfried Heinrich Ulau* (Leipzig: Eyssel, 1714). About Ulau little is known except that he defended his inaugural dissertation under Stahl in 1708.

[19] Georg Ernst Stahl, *Gründlicher physicalischer und medicinalischer Discurs eines berühmten Medici in Berlin von den warmen Bädern und Sauer-Brunnen* (Leipzig: Eyssel, 1716).

[20] Stahl, *Observationes clinico-practicae, worinnen gezeiget wird, wie ein Practicus die menschlichen Kranckheiten … heilen sole, Andere Auflage, … Ehemahln dem berühmten Kön. Preuss. Leib-Medico und Hof-Rath Herrn George Ernst Stahlen in einen Collegio Privatissimo Discurs-weise vorgetragen* (Leipzig: Eyssel, 1718).

[21] *Herrn George Ernst Stahls, berühmten Königl. Preußischen Leib-Medici und Hof-Raths, Gründliche Untersuchung der Kranckheiten, Welche bey einem jeglichen Alter des Menschen fürnemlich vorzukommen pflegen: aus dem Lateinischen übersetzt* (Leipzig: Eyssel, 1718); *Kurtze Untersuchung der Kranckheiten, welche bey dem kindlichen Alter des Menschen fürnemlich vorzukommen pflegen: aus dem Latein. übers.* (Leipzig: Eyssel, 1718); *Gründliche Abhandelung des Aderlassens, sowohl dessen Gebrauch und Mißbrauch, als auch dessen besondere Application auf dem Fusse und andern gewissen Theilen des Leibes: nebst einem ausführlichen Bericht, was von Aderlassen in hitzigen Fiebern zu halten sey; aus dem Lateinischen* (Leipzig: Eyssel, 1719); *Neu-verbesserte Lehre von den Temperamenten* (Leipzig: Eyssel, 1720).

[22] These two titles were published together sharing a title page: *Hn. George Ernst Stahls … Anweisung zur Metallurgie, oder der metallischen Schmeltz- und Probier-Kunst: Nebst dessen Einleitung zur Grund-Mixtion derer unterirdischen und metallischen Cörper. Alles mit gründlichen Rationibus, Demonstrationibus und Experimentis nach denen Becherischen Principiis ausgeführet* (Leipzig: Eyssel, 1720).

[23] Stahl, *Chymia rationalis*, 480.

[24] Stahl, *Verbesserung*, § 1–4, 18–22.
[25] Ibid, § 3, 20.
[26] Ibid, § 6, 22.
[27] Ibid, § 7, 23.
[28] Ibid, § 8, 23–24.
[29] The Latin original is *fundus*, and the German translation rendered it as *Capitalien*; ibid, § 13–14, 26–27.
[30] Fallopius, perhaps Gabriele Falloppio (1523–62), author of *Secreti diversi et miracolosi*, ed. Borgaruccio Borgarucci (Venice, 1664). Keßler may have been Thomas Kessle (fl. 1616–30), author of *Vierhundert ausserlesene chymische Process und Stücklein* (Strasbourg, 1632).
[31] Stahl, *Verbesserung*, § 17, 28–29.
[32] Pamela H. Smith elaborates this theme in *The Business of Alchemy* (Princeton: Princeton University Press, 1994).
[33] Stahl, *Verbesserung*, § 19–20, 29–30.
[34] Ibid, § 21, 30.
[35] Gottfried Wilhelm Leibniz and José Soares, *Novissima Sinica: historiam nostri temporis illustratura* (S.l.: s.n., 1697), and Christian Thomasius's review of *Confucius sinarum philosophus, sive scientia sinensi latiné exposita* (Paris, 1687), in his *Freimüthige, Lustige und Ernsthafte, Jedoch Vernunftmäßige Gedancken oder Monatsgeschpräche über allerhand, fürnehmlich aber neue Bücher* (Frankfurt: Athenäum Reprint, 1972), August 1689, 599–634.
[36] The Prorector is the faculty regent for the Rector, who, always a member of the royal family, was only the nominal head of the university.
[37] Johann Franz Buddeus, *Quaestionem politicam an alchemistae: Sint in Republica tolerandi* (Magdeburg: Christian Henckel, 1702). Following the contemporary usage, Buddeus meant polity by *Republica* in the title. For an analysis of Buddeus's thesis, see Ku-ming (Kevin) Chang, "Toleration of Alchemists as Political Question: Transmutation, Disputation and Early Modern Scholarship on Alchemy," *Ambix*, forthcoming.
[38] John Ferguson, *Bibliotheca Chemica: a Bibliography of Books on Alchemy, Chemistry, and Pharmaceutics*, 2 vols (Glasgow, 1906; reprint ed. Kila, MT: Kessinger Pub. Co., 1992), 2:300.
[39] Stahl, *Verbesserung*, 12.
[40] Michael Sendivogius, *Chymische Schrifften, darinnen gar deutlich von dem Ursprung, Bereit- und Vollendung des gebenedeiten Steins der Weisen gehandelt wird* (1718); Basil Valentine, *Via Veritatis, oder, Der einige Weg zur Wahrheit* (1718); Johann Joachim Becher, *Chymischer Rosen-Garten* (1717), *Opuscula Chymica rariora* (1719), and *Tripus hermeticus fatidicus* (1719).
[41] "De metallorum emendatione" was republished twice after its first appearance in 1703: in *Observationes physico-chymico-medicae curiosae* (Halle, 1709), a collection of Stahl's contributions to the *Observationes Hallensis*, and in *Opusculum Chymico-Physico-Medicum* (Magdeburg, 1715), a reprint of Stahl's early writings. Roth-Scholtz gives the wrong year (1702 instead of 1703) for the first publication; obviously he knew of no earlier publication of this text in the form of a dissertation in 1682, as Kopp and Schmieder suggest.
[42] Stahl, *Verbesserung*, 14.
[43] *G. E. Stahls Zufällige Gedancken und nützliche Bedencken über den Streit, von dem so genannten Sulphure* (Halle: Wäysenhauses, 1718).
[44] I cite the French translation that is available to me: "la substance qui donne à l'or sa couleur, sert en meme tems à les rendre plus déliés, à les rendre propres à occuper un moindre espace, sur-tout puisque l'argent lui-même, quand il a été transmuté en or par l'art, devroit surpasser la même masse plus petite dans le poids précedent." *Traité du soufre, ou remarques sur la dispute qui s'est élevée entre les Chymistes, au sujet du Soufre, tant commun, combustible ou volatile, que fixe* (Paris: Pierre-François Didot, le jeune, 1756), 130.
[45] Kopp, *Alchemie*, 71–73.
[46] "Ich kan ... aus manchfaltigen, zu meiner Erfahrung gelangten exempeln, versichern, daß ich die redlichste Ursachen von der Welt zu haben erdachte, zu glauben, daß die so grosse Leichtglaubigkeit, über dieser Fragen Bejahung, eine sehr ungegründete, und, wie so viel hundert exempel ausweisen, höchst anstößige und schädliche Sache sey; " Georg Ernst Stahl, *Ausführliche Betrachtung und zulänglicher Beweiß von den Saltzen*. (Halle: Wäysenhauses, 1723), § 36, 332–38, quotation on 335.
[47] "es ist nehmlich die allerzweiffelhaffteste Frage, ... ob auch wahr sey, oder jemals gewesen sey, daß ein Wesen von der Art und Krafft, wie die Tinctur beschrieben wird zu finden, oder zu verfertigen möglich

sey?" Georg Ernst Stahl, *Billig Bedencken, Erinnerung und Erläuterung über D.J. Bechers Natur-Kündigung der Metallen* (Frankfurt and Leipzig: Wolfgang Christoph Multz, 1723), 415.

[48] The treatises on sulfur and on salts were published by the press of *Waisenhaus*, which was directed by Stahl's Pietist colleagues. While *Waisenhaus* published a great number of inexpensive German Bibles for popular readers, it had also supported the academic publications of faculty members at Halle since its foundation in the last decade of the seventeenth century, and in particular it had published a number of Stahl's works, including his medical *magnum opus*, *Theoria medica vera* (1707–08). The publisher of the third work, Wolfgang Christophor Multz, based in Frankfurt am Main, seems to have published mainly academic medical monographs and dissertations.

[49] Stahl, "Bedencken von der Gold-Macherey," 3–4.

[50] Newman and Principe, "Alchemy vs. Chemistry," 39.

[51] Stahl, "Bedencken von der Gold-Macherey," 4.

[52] See, for example, Keith Tribe, "Cameralism and the Science of Government," *Journal of Modern History* 56, 1984, 263–84.

[53] Stahl, "Bedencken von der Gold-Macherey," 7–9.

[54] Ibid, 9.

[55] "welche durch dieser Art vergebliches Unterfangen, nimmer etwas wahres zum Stand oder Vorschein bringen;" ibid, 10.

[56] Ibid, 11–12.

[57] Ibid, 12–13.

[58] Ibid, 14–15.

[59] Ibid, 16.

[60] Ibid, 17–18. Although Stahl gave no detail about the reference to Orschall, it must be his *Sol sine Veste, oder Dreyssig Experimenta dem Gold seinen Purpur aufszuziehen* ... (Augsburg: Jacob Koppmayr, 1684). Stahl cited the story of Beuther told by Kunckel, which then is Kunckel, *Laboratorium Chymicum* (Hamburg and Leipzig, Samuel Heyl, 1716).

[61] Stahl, "Bedencken von der Gold-Macherey," 18–19.

[62] Ibid, 20.

[63] Ibid, 21–22.

[64] Ibid, 22.

[65] "[Ich] habe bey Wiederauflegung der Becherischen Chymischen *Concordanz*, auf Ersuchen des Hn. Verlegers, bißherige Bedencken, vornehmlich in der Absicht an den Tag legen wollen, daß man sich zum wenigsten desto mehrern vernünfftigen Nachdenckens und Überlegung befleißigen möchte, um nicht auf so wenig gegründete und bescheinigte Hoffnungen sich einzulassen ... Sondern vielmehr auf Natur, Vernunfft, und wahrer Kunst und Geschicklichkeit gemäße, Untersuchungen und Übungen, seine habende müßige Zeit, und überleye Kosten anwende. Als wodruch so wohl manche verständige Ergötzlichkeit, deutlichere Wissenschafft, und mäßige Vortheile und Nutzen, durch die wahre *Chymie*, und deren wohlgeübte Bearbeitung, erfolgen kan." Ibid, 23.

[66] Ibid.

[67] Ibid, 24.

[68] Ibid.

[69] See Johann Juncker, *Conspectus chemiae theoretico-practicae ... e dogmatibus Becheri et Stahlii potissimum explicantur*, 2 vols. (Magdeburg, 1730–38). As is clear from the title, even Juncker thought that the chemical teachings of Becher and Stahl harmonized so well that he discussed them together as if they formed a school. For an example among twentieth-century historians, see Rachel Lauden, "The Becher-Stahl School of Mineralogy and Cosmogony, 1700–1780," chapter 3 of *From Mineralogy to Geology: The Foundation of a Science, 1650–1830* (Chicago: University of Chicago Press, 1987).

[70] "Umständliche Erläuterung der Lehre vom Motu Tonico Vitali," in Georg Ernst Stahl, *Ausführliche Abhandlung von den Zufällen und Kranckheiten des Frauenzimmers* (Leipzig: Eyssel, 1724), 616–56.

[71] *D. Georg. Ern. Stahlii, consiliarii et archiatri regii Borussici primarii, Fundamenta chymico-pharmaceutica generalia: accessit manuductio ad enchirises artis pharmaceuticae specialis, cura Benjamin Roth-Scholtzii* (Herrnstadt: Samuel Roth-Scholtz, 1721). The publication of this text seems to have been a success, as it was

republished twice, though outside of Germany in Venice, in 1727 and 1741. Ferguson identifies Benjamin as Friedrich's pseudonym, and Samuel as Friedrich's third brother; see his *Bibliotheca chemica*, 2:297.

[72] *Die Alchemie des Andreas Libavius: Ein Lehrbuch der Chemie aus dem Jahre 1597. Zum ersten Mal in deutscher Uebersetzung* (Weinheim: Verlag Chemie, 1964).

[73] Christian Wolff, *Vernüfftige Gedancken von den Kräfften des menschlichen Verstandes und ihrem richtigen Gebrauche in Erkäntnis der Wahrheit, den Liebhabern der Wahrheit mitgetheilet* (Halle: Rengerische Buchhandlung, 1713).

[74] Goetze, *Scripta D. Georg. Ern. Stahlii*, 105–09.

[75] "ich in dem alten *Collegio chymico* von *anno 1684*, so letzthin von Herrn Lic. Carln ediret, in meinem damalen 25sten Jahr noch nicht so vollkommen von aller dergleichen Leichtgläubigkeit frey gewesen; wiewohl auch manches nicht ganz vergebens oder falsch sehn dürfte, wenn es bloß ad *veritatem physicam inveniendam* untersucht, nicht aber auf die thörichte transcendental-Hoffnung oder Einbildung der Goldmacherey angewendet würde." Juncker, *Conspectus*, 2, sigs. a4v–b1r.

JOHN C. POWERS

CHEMISTRY WITHOUT PRINCIPLES: HERMAN BOERHAAVE ON INSTRUMENTS AND ELEMENTS

INTRODUCTION

One of the more curious phenomena in the history of eighteenth-century chemistry was the reemergence of the four Aristotelian elements (fire, air, earth, and water) and of the alchemical notion of chemical "menstrua," and the recasting of these five as "instruments." These five instruments were defined as tools which the chemist utilized to instigate or prevent specific motions in matter during chemical operations. As such, the instruments and their specific properties occupied a prominent place in many pedagogical presentations of chemistry and were also the subject of theoretical discussion and experimental research. While some eighteenth-century chemists and more modern historians have referred to the "instruments" as "elements" (or "instrument-elements"), they were not elements in the traditional, Aristotelian sense.[1] They did not (as the Aristotelians held) enter into the composition of all bodies, and in fact, the extent to which the instruments, especially "air" and "fire," combined chemically with any other body was very much a topic for debate. In the seventeenth century, the earliest discussions by Daniel Sennert (c. 1620s) and the early university lectures of Georg Stahl (c. 1680s) clearly placed the instruments within the context of understanding the mechanisms or natural philosophy of chemical operations as distinct from problems of composition.[2] Derived from this context, the instruments in the eighteenth century represented a relatively novel shift in the interests of philosophically-minded chemists towards problems regarding the action and mechanisms of chemical operations.

The greatest proponent of the chemical instruments and font from which most eighteenth-century discussions stemmed was Herman Boerhaave (1668–1738), a lecturer and then professor of chemistry, botany, and medicine at the University of Leiden during the first three decades of the eighteenth century. In fact, the "instruments" were the center and foundation of his chemical theory. Boerhaave devoted almost half of his textbook, *Elementa chemiae* (1732) – over 600 pages – to examining the properties and uses of the "instruments." These "instruments" were a heterogeneous group, both ontologically and functionally. "Fire" was an imponderable, particulate fluid that acted as the cause of phenomena relating to heat, such as expansion, flame, and the destructive dissolution of bodies. "Air" was a ponderable fluid that exerted the pressure on bodies that was necessary, for example, during combustion, and also acted as a medium to contain vapors and "spirits." "Water" was both a ponderable

L. M. Principe (ed.), New Narratives in Eighteenth-Century Chemistry: Contributions from the First Francis Bacon Workshop, 21–23 April 2005, 45–61.

fluid and the most common solvent found in the chemist's laboratory. "Earth" was a simple, hard, insoluble body, which remained fixed in the fire, and whose function as an instrument was to "hold" volatile spirits, salts, and oils. Finally, the term "menstruum," did not refer to any specific species of body, but generally speaking to the power of any body to act as a solvent. Note, however, that Boerhaave's conception of a "menstruum" extended far beyond the modern notion of a "solvent." For example, Boerhaave included as menstrua the interactions and precipitations of salts in solution that modern chemists would identify as displacement reactions and later eighteenth-century chemists would call examples of "elective affinity." Similarly, Boerhaave included in this category the "power" of "fixed alkali" to absorb water from the air, literally melting if left unattended.[3] Within the logic of Boerhaave's chemistry, the instruments functioned as "natural" tools – this was the only property that unified the "instruments" as a class of chemical object.

Despite the ubiquity of the "instruments" in eighteenth-century chemistry, modern historical interpretations have tended to portray the idea of the "chemical instruments" as an obstacle to the conceptual development of chemistry. Henry Guerlac, for example, in his classic *Lavoisier – The Crucial Year* argued that the instrumental interpretation of "air" hampered chemists (at first) from seeing "air" as a possible reactant in chemical combinations – a necessary component of Lavoisier's theory of combustion. "Air" as an instrument was not (in Guerlac's words) a "chemical participant" in combustion, but rather a "mechanical" or "physical agent" merely keeping combustible particles in close proximity so that the "fire" may act on them effectively.[4] Guerlac's assertions on the physical character of the instruments have been misconstrued by subsequent historians and subsumed into a larger debate on the (at best) influence or (at worst) intrusion of "physics" into eighteenth-century chemistry.[5] Within this interpretive framework, the chemical instruments tend to be portrayed as a counter-productive intrusion of physics. Robert Siegfried, for example, has recently argued that Boerhaave's "mechanical" interpretation of the instruments was fundamentally not chemical:

> Mechanical views so dominated his conceptions that chemistry as such was left without any real identity. Indeed it is difficult at times to discern any chemistry at all in his conceptualization of the traditional four elements, which were assigned roles under the rubric "instruments," the agents by which chemical change is initiated and accomplished.[6]

Explicitly for Siegfried, and implicitly for Guerlac, what makes an account of material change "chemical" seems to be its focus on problems of composition. But as I stated earlier, the instruments were not devised to study composition.

My aim in this paper is to examine the case of Herman Boerhaave in order to suggest that the instruments constituted one solution to what many chemists considered to be a fundamental problem in traditional chymical approaches at the turn of the eighteenth century: claims about material composition based on analysis by fire. Many seventeenth-century chemists, for example Robert Boyle, began to doubt the efficacy of traditional fire analysis to decompose bodies into their principles or elements without altering the products of the analysis.[7] Following the work of Boyle, Boerhaave rejected the traditional chemical principles (i.e., salt, sulfur, mercury, and sometimes phlegm or water, and earth) on the grounds that they could not be extracted from bodies through chemical analysis and were, in fact, artifacts of both the chemist's operations and their undisciplined imagination. For Boerhaave the fundamental problem in chemistry was to determine what effects chemical operations actually had on matter and then to derive theoretical principles to guide the chemist's practice. He thus constructed his version of the chemical instruments as a means of scrutinizing the understanding of chemical species in terms of their component principles and shifting the focus of chemical philosophy toward examining the mechanisms of chemical operations. Note that by "mechanisms of chemical operations," I do not suggest an intrusion of "physics" or "mechanics" into chemistry. Rather, I argue that there was a shift in philosophical problems and methods within one, influential version of chemistry.[8]

Philosophical concerns, however, only reveal part of why Boerhaave adopted the instrument framework so enthusiastically. The pedagogical context and norms of the University of Leiden medical faculty served as both the motor and the model for Boerhaave's chemical lectures. Teaching in the medical faculty of a university, Boerhaave needed to provide his chemistry with a theoretical framework, but in the traditional didactic presentation of chemistry, the chemical principles served as the foundation for the discussion of the "theory" of chemistry. Thus, Boerhaave adopted and developed his account of the instruments to provide both a theoretical framework for examining chemical action and a "method" to organize diverse chemical phenomena.

This paper will examine how the instruments functioned, both as tools within the practice and philosophy of chemistry and as principles of organization within the structure of Boerhaave's chemical courses. First, I will discuss Boerhaave's adoption of Boyle's skepticism towards chemical principles and how he devised a research program to test the principles that he used in his own chemistry. Second, I will look at the pedagogical context of the instruments, especially how they acted as pedagogical *loci* that Boerhaave used to organize relevant chemical phenomena and to suggest problems for further research. Ultimately, I argue that the chemical instruments constituted a theoretical framework and method of organization for Boerhaave's chemistry, but not a dogmatic theory or "system." In fact, Boerhaave was well aware that the shift away from a principle-based chemical philosophy was a process, and the account of chemistry that appeared in his courses and in his *Elementa chemiae* was incomplete. Rather than limit the development of chemistry, as some modern historians have argued, Boerhaave intended his instrument framework to free chemistry from the dogmatism and conceptual confusion of the chemical principles in order to focus on *productive* problems.

BOERHAAVE'S CHEMICAL SKEPTICISM

Boerhaave's rejection of the chemical principles stemmed from the work of Robert Boyle. As has been well-documented by historians, the Leiden professor had a great admiration for Boyle, not only regarding his natural philosophy and programmatic investigation of chymical topics, but also regarding what Boerhaave took to be Boyle's religious devotion. Indeed, in Boerhaave's eyes, Boyle was a model Christian natural philosopher.[9] In his chemistry textbook, *Elementa chemiae* (1732), Boerhaave called Boyle "the great Master," and stated that he grounded his own program for chemistry on Boyle's principles.[10] Boerhaave adopted Boyle's skeptical attitude towards traditional chymical entities which led to a research program that tested chymical claims experimentally. Boerhaave's main source for this skepticism was, of course, the *Sceptical Chymist* (1661). Given that this text was a dialogue (and a disorganized one), Boerhaave also relied on Boyle's more programmatic texts, such as *The Producibleness of Chymical Principles* (first published as an appendix to the 1680 edition of the *Sceptical Chymist*) and the *Mechanical Origine or Production of Qualities* (1675), all of which he cited in his *Elementa chemiae* and earlier chemical lectures.[11] To understand why Boerhaave chose Boyle as his guide, however, one must examine the Leiden context a bit more closely. Boerhaave came of age as a chemist and physician during the 1690s, a time when Boyle's arguments against chymical principles were a topic of debate in Leiden and also a time when the Leiden medical faculty moved to expunge Cartesian and chymical medicine from the curriculum. Boyle's work, for Boerhaave, addressed both of these issues, and suggested a way to save chemistry in medicine.

Boerhaave first discovered the work of Robert Boyle while he was a student at the University of Leiden. As an undergraduate in the Liberal Arts faculty, Boerhaave studied mathematics and "physics" with Burchard de Volder (1643–1709), who also acted as his "promotor" during his graduation exercises in 1690. De Volder founded Leiden's *theatrum physicum* in 1676 and initiated a course in experimental physics based in large part on Boyle's pneumatic experiments with the air-pump.[12] When in 1691 Boerhaave began a plan of self-study in medicine to augment his study of theology, he had the opportunity to examine Boyle's published work closely. Boerhaave had obtained a position at the university library which allowed him a much greater access to books than the ordinary Leiden student. He took full advantage of this situation, reading the library's holdings (in chronological order, when possible) on medicine and chemistry, which included recent publications in English natural philosophy.[13]

Yet, if the salient features of Boyle's work escaped Boerhaave during his first reading, they were brought to the fore during his practical training in chemistry. Around 1692–93, Boerhaave worked in the laboratory of a local apothecary, David Stam (1633–1711). Stam himself had worked in the private laboratory of the illustrious Leiden professor, Franciscus Sylvius (De le Boë, 1614–72), and if his own statements are to be believed, he received a medical degree as well, although no record of this degree exists.[14] Stam prepared his only published work, an edition of John Francis Vigani's *Medulla Chymiae* (1693), while Boerhaave was an assistant in his laboratory. The *Medulla Chymiae* was a short primer on "recent" developments in chemistry, modeled on the traditional, didactic textbook structure. The book began with a definition and etymology for *chymia*, briefly examined the chemical principles, and then proceeded to discuss the properties and operations of chemical species, arranged in taxonomic order: salts, tinctures, plant extracts, animal extracts, and metals.[15] The *Medulla*, however, was not intended to be a comprehensive textbook. Vigani focused on relatively novel operations, and both his choice of operations and the framework with which he interpreted those operations was shaped by the work of Robert Boyle. In the first few pages of the *Medulla*, Vigani rejected the various systems of chemical principles, including the *tria prima* (salt, sulfur, and mercury), the Peripatetic elements, and the acid–alkali theory of Stam's mentor, Sylvius. Instead, Vigani posited that chemical change should be explained in terms of "atoms or corpuscles," citing Boyle's *Sceptical Chymist* as the authoritative source for this claim.[16] In response, Stam defended the work of his mentor in a long footnote – longer than Vigani's original argument – arguing that the *tria prima* were empirically demonstrated and universally accepted. Stam also cited his own list of authorities, such as Joan Baptista Van Helmont and Basil Valentine, to support his position.[17] Indeed, the whole of Stam's edition of the *Medulla* seemed to be a systematic refutation of Vigani's and Boyle's critique of traditional chymistry. In the preface to the book, Stam even asserted his commitment to the *tria prima* as the "most clear vocabulary of nature," and asserted their centrality to chemical investigation.[18]

Boerhaave ultimately sided with Boyle in this dispute, although his opinion was shaped as much by the philosophical climate in the Leiden medical faculty as by Boyle's arguments. Just as Boyle had critiqued the work of chemical physicians, like Franciscus Sylvius, by the 1690s the medical faculty at Leiden had also rejected the work of their most prominent professor. During Boerhaave's student days, two professors, Anton Nuck (1650–92) and Charles Drélincourt (1633–97) dominated the Leiden medical faculty, and both of these professors advocated an empirical approach to medicine grounded in anatomical observation and physiological experimentation.[19] They both rejected any approach to medicine which employed what they considered to be *a priori* assumptions rather than empirical observation to derive the mechanisms of the body and disease. This included both Cartesian medicine, which employed deductive reasoning to derive phenomena from first principles, as well as the Sylvian theory of acid and alkali. Note, however, that neither Nuck nor Drélincourt banished mechanism or even chemical methods completely from their medicine. Both portrayed the body very much as Boerhaave would do later: as a mechanical object to be understood in terms of the physical and chemical properties of its solid parts and fluids.[20] But both Nuck and Drélincourt placed the empirical work of the anatomist and physiologist philosophically prior to theoretical explanation, and they worked to remove *speculative theorizing* from the medical teaching at Leiden. Drélincourt, in fact, published a critique of Sylvian chemical medicine (under a pseudonym) in 1668 while both he and Sylvius were on the medical faculty. Later in 1690, Nuck penned a letter to the university curators which induced them to block Jacob le Mort, a local apothecary and chemical lecturer, and Cornelius Bronktoe, and a medical lecturer and Leiden graduate, both Cartesians, from giving private lectures at the university.[21]

When Boerhaave was appointed to the medical faculty as a lecturer in 1701, he continued to promote the empirical, anti-speculative ideology of Nuck and Drélincourt. This ideology was evident in Boerhaave's very first academic oration at Leiden (1701) entitled, "On Recommending the Study of Hippocrates," in which he portrayed the Greek Father of Medicine as a strict and prudent empiricist. According to Boerhaave, Hippocrates never taught a remedy to his students or recorded its preparation in writing until "he had proved [it] to be successful in a thousand cases," and even then he presented the remedy with "cautions and warnings" in order to prevent the inexperienced physician from misusing it.[22] Boerhaave contrasted this model of medical practice with that of contemporary Cartesian physicians. In the closing lines of the oration, he argued that the Cartesians,

> prefer to take general principles of nature, like matter, movement, and the shape of corpuscles, as their starting point – *a priori*, as they call it – in demonstrating the essence of health, of diseases, and remedies. Not at all embarrassed by a multitude of unproven assumptions, a lack of factual data, they set up a vague hypothesis; then, having posited this, they draw some rather sweeping conclusions via analytical thinking and using seemingly plausible reasoning; afterwards they apply this to the facts and do not hesitate to invent rules for healing based on this kind of ingenuity. But now it has become evident that they are disappointed in their ambitious expectations when they try to use these figments of their imaginations in practical medicine, much to the detriment of their patients.[23]

Although Boerhaave identified the Cartesians as the greatest contemporary threat to medicine, he reserved similar criticism for Sylvian chemical medicine, which he described as composed of "fables" and practiced by "quacks" and "mountebanks."[24]

Boerhaave extended his critique of speculative theorizing to chemistry as a part of his program to reform medicine. In 1718, after his appointment to the Chair of Chemistry at Leiden, he presented another oration titled, "On Chemistry Expunging its Errors." Here he argued that one of the major errors that "modern" chemistry had expunged was Sylvius's theory of acid and alkali and its mis-application in medicine. Like the Cartesian physicians, Sylvian chemical medicine was undermined by its faulty methodology. He asserted that Sylvius's whole system was based on the speculative assumption that relatively simple chemical reactions that one performed outside the body could be models for the body's physiological mechanisms. Boerhaave presented numerous examples of how this analogy worked in practice. For instance, since distilled acids mixed with essential oils generated a "fervid" heat, the chemical physicians argued that the mixing of the acid found in chyle with the "balsam of blood" (an oil) accounted for the warmth of the body. Similarly, the cause of "burning fevers" was attributed to stronger mixtures of these two substances.[25] Boerhaave rejected this reasoning by analogy because it assumed one simple rule (the antipathy of acids and oils) to explain all relevant phenomena and, by doing so, did not represent the complexity of nature. To emphasize the absurdly simplistic nature of chemical medicine, Boerhaave commented in his oration that one may learn the basic precepts of this art in just one hour. One need only to learn a little about the natures of "acid" and "alkali" and to realize that the whole endeavor involved achieving an "equilibrium of forces" between the two.[26]

Boerhaave read Robert Boyle's work as being philosophically in agreement with the Leiden medical faculty's rejection of speculative theorizing. Boyle deployed experiments systematically to examine chemical claims in order to refute those which could not withstand empirical scrutiny. In this regard, Boyle's skeptical approach to chemistry was a useful tool for Boerhaave's program to expunge chemical "errors" from the medical curriculum. In the *Elementa chemiae*, Boerhaave praised Boyle for his use of chemistry "to discover and correct those errors, which some whimsical dabblers in chemistry had introduced into Medicine."[27] Boerhaave was quick to point out, however, that chemistry did have a proper place in medicine, contributing to all of the major subjects of the field. The key was finding the proper philosophical and methodological approach. Nuck and Drélincourt, both being primarily anatomists and physiologists, had little help to offer here. Boyle, however, presented a program of experimental investigation designed to test theoretical claims and which promoted the same empiricism found in Boerhaave's medicine.

Taking his cue from Boyle, Boerhaave treated the chemical principles in the same way that he treated Sylvius's acid and alkali theory: as theoretical speculations. Boerhaave accepted Boyle's arguments found in the *Sceptical Chymist* and elsewhere that asserted that the so-called principles or elements chemists claimed to isolate through their analyses were, in fact, not the primary constituents of bodies, but rather were artifacts of the chemists' operations themselves. He presented this argument as a thesis in his very first chemistry course (1702) as follows:

> The products of separation (i.e., chemical analysis) can neither be resolved into similar parts or elements, nor are they always produced by the separation alone without the mutation of their parts.[28]

The lecture notes in which this "thesis" is recorded do not indicate how Boerhaave may have elaborated on and interpreted this statement for his students. He clarified his position in much greater detail in the *Elementa chemiae*. Regarding the operations of chemistry deployed for analysis, Boerhaave argued that the same "powers" which disunite bodies "may produce in them likewise a great alteration, and we shall fall into error if we suppose that the [original] compound bodies in reality do contain these very elements."[29] The chemist's fire was just as likely to rearrange as to separate the "corpuscles" of compound matter, and thereby produce new effects "which never [presented] themselves by any effect in the bodies while they were entire." Thus, Boerhaave concluded, the "chemists" are mistaken when they claim to produce the "first elements of bodies" and "think they can determine the nature of compounds" from these elements. Indeed, the only entities which Boerhaave argued may rightly be called "elements" are the "corpuscles" or "atoms" which compose every body. But, he asserted, as a practical matter these corpuscles can never be collected in a pure state by the chemical art.[30]

Despite this seeming rejection of the chemical "elements," Boerhaave found that he could not remove all the hypothetical principles that shaped earlier, chymical explanations of phenomena central to chymical practice. The "Theory" section of the *Elementa chemiae* was full of entities that he referred to as "principles" or "elements." In fact, Boerhaave presented a list of "chemists's elements" immediately following his general critique of those same elements. Included on the list were four of the "instruments" – fire, air, water, and earth – and three others: the "Alcohol of wine, Mercury (of metals), and the *Spiritus Rector* of every body."[31] But Boerhaave then elaborated on what he meant by "element." He argued that while they "do appear exceedingly subtle and durable," none of these entities could be obtained in pure state by chemical methods. Thus, one could not posit with certainty that, according to the traditional definition of an "element," any of these entities were truly the constituents of bodies. For the four "instruments" on the list, this was primarily an academic point, since in Boerhaave's chemistry they functioned as tools and at first did not play an important role as constituents of bodies. For the other three "elements," however, the problem of analysis posed a larger, philosophical problem. In Boerhaave's chemistry and indeed in the traditional chymistry from which they derived, these "elements" conferred specific properties to the substances into which they were compounded. For example, "alcohol of wine" was the purest form of Boerhaave's "*pabulum ignis*" – his principle of inflammability – that was obtainable through chemical means. Derived from the *tria prima*'s "sulfur" principle, the *pabulum ignis* functioned in Boerhaave's chemistry as the material cause of combustion. Bodies containing *pabulum ignis* would burn if given the proper application of fire; those without it would not.[32]

In an attempt to place these "elements" on a solid philosophical foundation, Boerhaave conducted a series of investigations modeled on Boyle's "skeptical" examination of chymists' claims, and designed to isolate and examine each of these "elements," treating them like ordinary chemical species. As described in the *Elementa chemiae*, he began his investigation with a consideration of the products one obtained through the dry distillation and calcination of vegetable matter, asking which parts of this analysis were combustible. After reviewing each product – water, spirits, oils, smoke, salts, etc. – Boerhaave concluded that the combustible products were the oily ones.[33] Not all oils, however, burned easily. Oil of turpentine, for example, was difficult to burn, unless it had been "attenuated" though art: successive distillations, putrefaction, or fermentation. For Boerhaave, the most attenuated oil, and thus, the purest form of the *pabulum ignis*, was alcohol of wine, which was the product of both fermentation and distillation.[34] In the hope of observing or collecting the *pabulum ignis* itself, Boerhaave conducted a series of experiments in which he burned the purest alcohol that he could make under a bell-jar designed to collect the products of combustion. Despite his efforts, his experiments produced only water with no other detectible product. After a protracted discussion in the text, Boerhaave concluded that the *pabulum ignis* must have been fixed in the water (thus, alcohol is water combined with the *pabulum*), and released by the action of the fire. The *pabulum* itself must be "so vastly fine, that it is dissipated into the chaos of the Atmosphere, and goes beyond the reach of our senses."[35] Boerhaave undertook similar investigations for his *Spiritus rector* and his Mercury, both presenting similar, inconclusive results.[36]

From these examples, we can begin to understand the aim of Boerhaave's chemistry and the function of the chemical "instruments" within it. For Boerhaave, chemical operations could not be deployed reliably to make claims regarding the composition of compound bodies. Because of this problem Boerhaave saw the need, as he stated in the *Elementa*, to "fix some sure limits to our Art, which we must not exceed if we would avoid mistakes, and come at the truth."[37] First among these limits was determining in a rigorous manner the effects that chemical operations generated in matter. If chemical analysis could not produce "elements" or "principles," what effects did it produce, and could these effects be understood systematically? One method of answering this question was simply to examine and organize individual chemical operations. The result of this endeavor, exemplified in the second volume of the *Elementa*, resembled the traditional, didactic presentation of chemistry found in the chemical artisans' textbooks. Another method was to examine the "tools" that chemists employed in conducting their operations. This explains why Boerhaave devoted so much attention to the chemical "instruments." By examining the properties, "powers," and effects of (for example) "fire" in all of its guises, the chemist would better be able to control and predict the effects of "fire" in his operations. Thus, Boerhaave's theory of chemistry focused on understanding the effects of the instruments on matter, not on the elemental composition of bodies.

INSTRUMENTS OF CHEMISTRY

Boerhaave's discussion of the chemical instruments dominates the "Theory" section of the *Elementa chemiae*. They were the main vehicles through which he examined the effects of the chemical art on matter. In effect, Boerhaave's chemical theory presented the outline of a research program designed to examine the natural properties of the instruments and how they interact with and through chemical species. Why Boerhaave adopted his instruments approach to chemical theory, however, is only partially explained by his chemical skepticism. Boerhaave developed his chemical system within a pedagogical context. He taught chemistry at the University of Leiden and, following the pedagogical norms of the medical faculty, he needed to devise both a theoretical presentation for chemistry and a method to organize chemical phenomena. The instruments provided both of these things. Within his courses and textbook, the instruments acted as pedagogical *loci*, organizing chemical phenomena for didactic purposes. They also functioned as the focus for Boerhaave's experimental philosophy of chemistry, suggesting a research program through which the chemist could derive principles of chemical action.

When Boerhaave began to teach chemistry at Leiden, his first challenge was to design a course that followed the pedagogical norms of the university. In January 1702, Boerhaave presented his first chemical lecture course in the medical faculty. In the following winter term, beginning in October 1702, he offered his lecture course again along with a course of chemical demonstrations.[38] These two courses reflected the standard curricular approach to medicine at the University of Leiden by dividing the field into "theory" (the lecture course) and "practice" (the demonstrations course). In the medical faculty, students ideally studied the theoretical principles of medicine in the "Institutiones medicae" course before applying those principles in their "praxis medica" course and, eventually, through clinical instruction.[39] Dividing chemistry according to the curricular requirements of the university, however, posed several challenges for Boerhaave. Traditional, didactic presentations of chemistry as exemplified by contemporaneous chemical textbooks and in the chemistry courses given in Leiden by previous lecturers, focused overwhelmingly on describing chemical recipes and operations and their uses. While Boerhaave employed these "courses" and their method of organization effectively in his demonstration course, he faced a problem in his lecture or "theoria" course. Most textbooks devoted only a few pages at the beginning of the text to a theory of chemistry. These "theory" sections tended to focus on definitions and etymologies of chemical terms and the discussion of whatever system of chemical principles (i.e., the *tria prima* and its variants) the author preferred.[40] By 1702, however, the notion of a system of chemical principles acting as the theoretical foundation for his chemistry was anathema to Boerhaave. Thus, he had to find a new theoretical structure around which he could shape his lecture course.

He found the roots of his new structure in a series of academic dissertations published by the Leipzig medical professor, Johannes Bohn (1640–1718). Bohn's *Dissertationes chymico-physicae* (1685) consisted of 15 sets of dissertations on "chemical" topics, each containing 20 or 30 numbered theses. Each thesis ranged from a few lines to an entire page in length and presented a definition or philosophical proposition relating to the nature of the chemical art, a "chemical" entity, or a phenomenon. Bohn had written the theses for his students to defend in the academic exercises conducted in the University of Leipzig's medical faculty. In these exercises, the aim of the student *defendens* was to argue for the logical import of each thesis, thereby demonstrating that he understood the philosophical principles behind the thesis. As molded by this context, Bohn's theses focused on definition, philosophical argument, and theoretical explanation rather than on chemical recipes or practice *per se*.[41]

In Bohn, Boerhaave found a model for his lecture course. Like other Leiden professors, Boerhaave composed his medical, and now chemical courses as theses in imitation of the academic exercises through which students were examined. In addition, Boerhaave saw Bohn as a philosophical ally. Bohn was a mechanist, but he did not endorse the "speculative" theories of the Cartesians.[42] Most importantly, he too was a great admirer of Robert Boyle. In the first dissertation on the "dissolution" of bodies, Bohn addressed the question of whether one can resolve a "mixt" into its component principles or elements. After reviewing the claims for such resolutions, such as extraction of the mercury of metals and the "principles" of plants, Bohn concluded that one could not "analyze" bodies into their elements. For this reason, Bohn asserted that his chemistry will avoid (*eludunt*) the claims of the "philosophers of fire" and "peripatetics" regarding the elements. For textual support of this claim, he cited that "Illustrissimus Philosophus Angliae," Robert Boyle, in his *Sceptical Chymist*.[43]

Bohn's *Dissertationes* also presented an approach to the chemical instruments that Boerhaave could adopt to his own purpose. At the beginning of the third dissertation, on "Fire," Bohn described the furnaces, vessels, filters, etc. that chemists employed in their work as the "instruments" through which they enacted the dissolution and concretion of bodies.[44] In addition to these artificial instruments, there were "natural" instruments – fire, air, water, earth and chemical *menstrua* – which acted as media between the chemist and the chemical species he wished to manipulate. The natural instruments were able to effect changes in chemical species due to their natural properties and "powers," which any competent chemist must understand so that he could establish proper rules and precepts for conducting chemical operations. In total, Bohn devoted four dissertations to describing the properties of the natural instruments. Each of the remaining nine dissertations examined a specific class of operation (i.e., digestion, calcination, fermentation, etc.), and Bohn argued how each operation utilized one or more of the instruments to generate its effects.

Boerhaave composed an outline from Bohn's *Dissertationes* that contained all of the theoretical points and observations (drawn from Bohn) that he would later incorporate into his chemical lectures. Like Bohn's dissertations, Boerhaave's outline was organized as a series of numbered theses, each consisting of a definition or statement of theory regarding a phenomenon or concept.[45] In effect, Bohn provided Boerhaave with a body of concepts and relevant facts from which to construct his lecture course. Note that Boerhaave did not simply copy Bohn's *Dissertationes* verbatim. He molded Bohn's work to fit his own program, removing or adding material to suit his own needs. For example, within his discussion of fire, Bohn discussed in detail John Mayow's theory of "aerial-niter": a principle of inflammability present in the air that explains the necessity of air in combustion.[46] Boerhaave, however, omitted the "aerial-nitre" from both his outline and from his chemical courses, preferring initially to ascribe to air a purely mechanical function in combustion.[47] Nevertheless, when Boerhaave drafted his first chemical lecture-course, he wrote it as a series of 264 theses, the theoretical center of which – the chemical instruments – mirrored his outline of Bohn's *Dissertationes*.[48]

Perhaps the most important thing with which Bohn provided Boerhaave was a "method," via the chemical instruments, to organize chemical phenomena according to theoretical principles. By "method" I mean a pedagogical method deployed to order topics in an academic field through taxonomic division and explication.[49] Each instrument functioned as a taxonomic category, explaining a set of phenomena that Boerhaave deemed relevant to chemical practice. Within each instrument-category, Boerhaave further divided phenomena according to the natural "principle" that explained them, usually a specific form of interaction between a species of matter and the chemical instrument. For example, in his first lecture course under the topic, "fire," Boerhaave differentiated two general types of fiery phenomena: "burning" (*urens*) fire, which ultimately focused on combustion, and "shining" (*lucens*) fire, which he associated with light. Under the heading of "burning" fire, Boerhaave posited the theory that all phenomena associated with "fire" generated heat through motion caused by "attrition"; that is, the agitation or "rubbing" (*atterens*) of particles of fire by matter and vice versa. He presented four types of agitation, beginning with simple mechanical action: the concussion of a hammer on metal, which generates heat in both the hammer and the metal. Using this straightforward, empirical example, Boerhaave established a simple connection between heat and mechanical motion. He then proceeded to more complex phenomena, such as heat generated through the violent reaction of oil of vitriol with vegetable oil and the combustion of phosphorus.[50] In each of these cases, the heat generated was explained in terms of the mechanical "attrition" between particles of "fire" and the particles of other bodies. This move from simple, empirical examples to complex ones, showing how each could be understood as the result of the same "principles," was the foundation of Boerhaave's "method."

Boerhaave's use of this method (simple to complex) to elucidate theory signaled a fundamental shift in the pedagogical practices of chemistry. In effect, he deflected the emphasis of his chemical courses away from the properties of chemical species and their operations to the theoretical principles through which operations may be understood philosophically. Boerhaave's philosophical chemistry received its full expression in a ten-year series of lecture courses on the "instruments" that he presented in succession from 1718 through 1728.[51] These lectures, which collectively I have called the "Instruments Course," incorporate material from his year-long lecture course, but also significantly re-work and expand much of this material.[52] A good portion of the Instruments Course offered theses on the instruments like his older course, but for many topics, Boerhaave opted to present phenomena and principles through systematic demonstration-experiments. Typically, he designed these demonstrations to show how the theoretical principles which guided the behavior and chemist's use of the instruments could be established experimentally. Like the rest of his course, these demonstrations were presented according to Boerhaave's method. For any topic, the first experiments exhibited simple, basic phenomena, from which Boerhaave constructed more complex experiments, ultimately exhibiting and explaining phenomena which directly effected chemical operations.

A good example of Boerhaave's use of demonstration-experiments in his Instruments Course was found at the beginning of his lectures on "fire." Boerhaave began this course by positing the question of how one may reliably determine the presence of fire in a body; that is, what is the "sign of fire"? After suggesting and discarding several possibilities, such a flame, the sensation of heat, etc., he settled on volumetric expansion as the sign. He then presented a series of experiments which demonstrated the relative expansion of various bodies – an iron rod, an iron circle, and various fluids – when heated in the fire, arguing that for each species of body, the amount of expansion was proportional to the relative amount of fire each body had absorbed. These experiments culminated with the demonstration of a Fahrenheit thermometer, which Boerhaave offered as the ideal tool for measuring fire in bodies.[53] In subsequent lectures Boerhaave demonstrated how the thermometer could be used to examine other phenomena relating to fire. In one series of demonstrations, he exhibited to his students how chemical species, when mixed together, may generate or lose heat and how the thermometer may be employed to register the "degrees of heat" evolved though this interaction. He began by mixing distilled water with spirit of wine, but later progressed to the mixing of other liquids, the dissolving of various salts in water, the mixing of acids and alkalis, and the dissolution of iron filings in oil of vitriol.[54]

Through these series of demonstration-experiments, Boerhaave was not only conveying to his students relevant chemical phenomena and principles, but he also outlined a research program in chemistry centered around the chemical instruments. At the conclusion of this thermometry experiments described above, Boerhaave stated this aim overtly. He maintained that the goal of his experimental approach was to generate a "complete and certain history of Heat." He suggested to his students that all "simple bodies" from the three kingdoms of nature should be examined in this way, first by mixing them with other bodies of their "class," and then with bodies of other classes. Ultimately, he recommended to them that they each obtain one of Fahrenheit's thermometers and continue this work themselves.[55] Boerhaave made similar suggestions throughout his Instruments Course and, later, in the *Elementa chemiae*, identifying problems or new areas of investigation for his students and readers.

Boerhaave's version of the chemical instruments was a product of the university context. While he acknowledged his debt to the artisanal chemistry of the didactic textbooks, Boerhaave's empirical philosophy, skeptical attitude, and focus on establishing theoretical principles was derived from academic and philosophical traditions. The audience for his courses were medical students, whose future status in the traditional medical hierarchy and republic of letters over that of artisan-practitioners and "empirics" depended upon their command of philosophical knowledge. The curriculum of the Leiden medical faculty was structured to train physicians for this role, and Boerhaave designed his chemistry courses to accommodate this standard.

CONCLUDING THOUGHTS ON "AIR"

In concluding I would like to return for a moment to the issue brought up by Guerlac concerning the instrument "air" within the context of Boerhaave's overall philosophical attitude towards chemistry. Recall that Guerlac suggested that the status of "air" as an unreactive "instrument" prevented chemists from discovering its ability to combine chemically with other bodies. In response, I argue that Boerhaave, in fact, came to the realization that air may be fixed in bodies as the result of his own experiments in preparation for his Instruments Course. The notes from his first chemical course in 1702 indicate that Boerhaave did initially maintain that air was unreactive. But when Boerhaave began to work out the demonstration-experiments for his lectures on "air" in 1722, he had a change of heart. He had designed a series of experiments initially intended to show how air dissolved in liquids could be removed through mechanical means, such as heat, chemical agitation, or exposing the liquid to a vacuum. In one experiment, Boerhaave placed a flask of water in an air-pump, evacuating the receiver and exhibiting the effervescence of the air from the water as he reduced the air pressure. He then mixed one-half dram of crabs' eyes (an alkali) with one and a half ounces of "ardent" vinegar, explaining (in his lecture) that the effervescence generated by this reaction was caused by the dissolved air being expelled through the violent motion of the acid and alkali. He attempted another version of this experiment with oil of vitriol and oil of tartar, but he let the liquids stand *in vacuo* for 25 hours, expecting that there would thereafter be little or no effervescence. To his surprise, mixing the two liquids generated a great deal of "elastic air." When he presented this demonstration in the Instruments Course, he explained that "during the effervescence as the attracted bodies rush together in mutual embrace, air that was intimately united with the elements of those bodies is expelled, making [elastic] air."[56] He further built upon this idea in the *Elementa chemiae*, where he reinterpreted many of his air-generating experiments from his earlier lectures on "air" in terms of the release of fixed air as opposed to air forced out of solution. He also added several more examples of air "united with other non-aerial particles." Perhaps emboldened by Stephen Hales's *Vegetable Staticks*, which he cited as the authoritative source on fixed air, he asserted: "elastic air occurs as a pretty considerable and remarkable constituent part in the composition of almost all kinds of bodies."[57]

The example of Boerhaave's changing views on fixed air undermines the historical interpretation of the instruments as a dogmatic "theory" which prejudices its adherents to see the instruments only as mechanical agents. As I have argued, Boerhaave adopted the instruments approach as an alternative to the dogmatism that he saw in the theories and methods of the Cartesian physicians, the practitioners of Sylvian chemical medicine, and the "vulgar chymists" criticized by Robert Boyle. In fact, Boerhaave was fully aware that his chemistry was incomplete and imperfect. He composed many sections of the *Elementa chemiae* in a manner to encourage further work that would clarify or correct problems, omissions, or inconsistencies that Boerhaave himself recognized. One of the most striking examples of this occurred during Boerhaave's examination of "air." At the end of a discussion on the necessity of air for life, he stated that there was a "certain hidden virtue" in air "which cannot be accounted for from all the properties of air, which have been hitherto discovered." The alchemist Michael Sendivogius, he asserted, called it the "food of life." While Boerhaave suggested that there must be some "physical cause" for this property of air, he remained agnostic on the question. Instead, he encouragingly asserted: "Happy is the person that shall happen to discover it."[58]

While the dominant historiography of chemistry has portrayed the notion of the chemical instruments as a theory that needed to be discarded by the end of the eighteenth century, I would like to focus on its positive aspects. Within Boerhaave's framework, the instrument approach called attention to major problems in eighteenth-century chemistry, such as the roles of air, fire, and menstrua in chemical operations, and how one goes about understanding those roles philosophically. Boerhaave's aim was not to create a dogmatic theory or system of chemistry, but rather to point out the errors of previous systems. Through the chemical instruments, Boerhaave replaced the older systems with a method of organizing and understanding chemical phenomena that sought to bring conceptual order to the field as well as to encourage further investigation.

NOTES

1 On the notion of instrument-elements, see Rosaleen Love, "Herman Boerhaave and the Element-Instrument Concept of Fire," *Annals of Science* 31, 1974, 547–59; David Oldroyd, "An Examination of G. E. Stahl's *Philosophical Principles of Universal Chemistry*," *Ambix* 20, 1973, 36–52.

2 Cf. Daniel Sennert, *De chymicorum cum Aristotelicis et Galenicis consensu ac dissensu liber* (Wittenberg, 1629). Stahl's lectures at the University of Jena from the early 1680s were published by a former student as Georg Ernst Stahl, *Fundamenta chymiae dogmaticae et experimentalis* (Nuremberg, 1723), and in English as *Philosophical Principles of Universal Chemistry*, trans. Peter Shaw (London, 1730).

3 Hermann Boerhaave, *Elementa chemiae*, 2 vols. (Leiden, 1732). I have used Boerhaave, *Elements of Chemistry*, ed. Timothy Dallowe, 2 vols. (London, 1735), 1:78–500. On Boerhaave's interest in fixed alkali

see Boerhaave, *Elements*, 440–62. On elective affinity, see Alistair Duncan, *Laws and Order in Eighteenth-Century Chemistry* (Oxford: Clarendon Press, 1996).

[4] Henry Guerlac, *Lavoisier – The Crucial Year: The Background and Origin of His First Experiments on Combustion in 1772* (Ithaca, NY: Cornell University Press, 1961), 20–24 and *passim*.

[5] On the shaping of chemistry by "physics," see, for example, Arthur Donovan, *Antoine Lavoisier: Science, Administration, and Revolution* (Cambridge, MA: Blackwell, 1993), esp. 45–73; "Lavoisier and the Origins of Modern Chemistry," *Osiris* 4, 1988, 214–32; Evan M. Melhado, "Chemistry, Physics, and the Chemical Revolution," *Isis* 76, 1985, 195–211; Maurice Crosland, "Chemistry and the Chemical Revolution," 389–416, on 391 in *The Ferment of Knowledge: Studies in the Historiography of Eighteenth-Century Science*, eds. G. S. Rousseau and Roy Porter (Cambridge: Cambridge University Press, 1980); Henry Guerlac, "Chemistry as a Branch of Physics: Laplace's Collaboration with Lavoisier," *Historical Studies in the Physical Sciences* 7, 1975, 193–276. For a brief overview and critique of this interpretation of eighteenth-century chemistry, see Carlton Perrin, "Chemistry as a Peer of Physics: A Response to Donovan and Melhado on Lavoisier," *Isis* 81, 1990, 259–70; Frederic Lawrence Holmes, *Eighteenth-Century Chemistry as an Investigative Enterprise* (Berkeley, CA: Office for the History of Science and Technology, 1989), 103–11.

[6] Robert Siegfried, *From Elements to Atoms: A History of Chemical Composition* (Philadelphia: American Philosophical Society, 2002), 129.

[7] Allen G. Debus, "Fire Analysis and the Elements in the Sixteenth and Seventeenth Centuries," *Annals of Science* 23, 1967, 128–47.

[8] What kind of phenomenon or theoretical framework defined "chemistry" (or "physics") was a central issue in the eighteenth century. See, for example, on chemistry Christoph Meinel, "Theory or Practice? The Eighteenth-Century Debate on the Scientific Status of Chemistry," *Ambix* 30, 1983, 121–32; J. R. R. Christie and J. V. Golinski, "The Spreading of the Word: New Directions in the Historiography of Chemistry," *History of Science* 20, 1982, 235–66; J. B. Gough, "Lavoisier and the Fulfillment of the Stahlian Revolution," in *The Chemical Revolution: Essays in Reinterpretation*, ed. Arthur Donovan, *Osiris* 4, 1988, 15–33; Martin Fichman, "French Stahlianism and Chemical Studies of Air, 1750–1770," *Ambix* 18, 1971, 94–122. On the fluid context of "physics," see John L. Heilbron, *Elements of Early Modern Physics* (Berkeley, CA: University of California Press, 1982); Simon Schaffer, "Natural Philosophy," 55–92 in *The Ferment of Knowledge*.

[9] Rina Knoeff, *Herman Boerhaave (1668–1738): Calvinist Chemist and Physician* (Amsterdam: Koninklijke Nederlandse Akademie van Wetenschappen, 2002), 115–16; Andrew Cunningham, "Medicine to Calm the Mind: Boerhaave's Medical System and Why It Was Adopted in Edinburgh," 40–66, on 48 in *The Medical Enlightenment of the Eighteenth Century*, eds. Andrew Cunningham and Roger French (Cambridge: Cambridge University Press, 1990).

[10] Boerhaave, *Elements*, 52 and 56.

[11] Robert Boyle, *Sceptical Chymist* (London, 1661); *The Producibleness of Chymical Principles* (Oxford, 1680); *Experiments, Notes, &c. About the Mechanical Origine or Production of Divers Particular Qualities* (London, 1675). On Boyle's *Sceptical Chymist* and his critique of chymical principles, see Lawrence M. Principe, *The Aspiring Adept: Robert Boyle and His Alchemical Quest* (Princeton: Princeton University Press, 1998), 27–62 and Debus, "Fire Analysis."

[12] C. De Pater, "Experimental Physics," 309–27 in *Leiden University in the Seventeenth Century*, eds. T. Th. Lunsingh Scheurleer and G. H. M. Posthumus Meyjes (Leiden: Universitaire Pers Leiden/Brill, 1975); Gerhard Wiesenfeldt, *Leerer Raum in Minervas Haus: Experimentelle Naturlehre an der Universität Leiden, 1675–1715* (Amsterdam: Koninklijke Nederlandse Akademie van Wetenschappen, 2002); Burchard de Volder, *Quaestiones academicae de aëris gravitate* (Middelburg, 1681).

[13] Boerhaave, "Commentariolus," 379–81 in G. A. Lindeboom, *Herman Boerhaave: The Man and His Work* (London: Metheren, 1968).

[14] On Stam, see G. A. Lindeboom, "David en Nicolaas Stam, apothekers te Leiden," *Pharmaceutisch Weekblad* 108, 1973, 153–60.

[15] Giovanni Francesco Vigani, *Medulla Chymiae*, ed. David Stam (Leiden, 1693). On Vigani, see L. J. M. Coleby, "John Francis Vigani," *Annals of Science* 8, 1952, 46–60; Anita Guerrini, "Chemistry Teaching at Oxford and Cambridge, Circa 1700," 183–99, on 186–88 in *Alchemy and Chemistry in the Seventeenth and Eighteenth Century*, eds. Piyo Rattansi and Antonio Clericuzio (Dordrecht: Kluwer, 1994); John R. Partington, *A History of Chemistry*, 4 vols. (New York: St. Martin's Press, 1962), 2:686–87.

[16] Vigani, *Medulla*, 11–13.

[17] Stam, in Vigani, *Medulla*, 13–18. Note that Van Helmont's view of the chymical principles was much more complex and ambivalent than as presented by Stam. Cf. William R. Newman, *Gehennical Fire: The Lives of George Starkey, and American Alchemist in the Scientific Revolution* (Cambridge: Harvard University Press, 1994), 110–14.

[18] Stam, "Proemium" in Vigani, *Medulla*, 1–9, on 1: "… hoc Naturae vocabulum paulo clarius illustrem."

[19] On Drélincourt's and Nuck's experimentalism, see G. A. Lindeboom, "Frog and Dog: Physiological Experiments at Leiden during the Seventeenth Century," 179–93, on 289–91 in *Leiden University in the Seventeenth Century*. On anatomical empiricism, see Andrew Wear, "William Harvey and the 'Way of the Anatomists,'" *History of Science* 21, 1983, 223–49.

[20] On Nuck's and Drélincourt's mechanism, see Antonie M. Luyendijk-Elshout, "*Oeconomia Animalis*, Pores, and Particles," 295–307, on 302–03 in *Leiden University in the Seventeenth Century*; Edward G. Ruestow, "The Rise of the Doctrine of Glandular Secretion in the Netherlands," *Journal of the History of Medicine* 35, 1980, 265–87, on 266–67.

[21] On Drélincourt's critique, see Lindeboom, "Frog and Dog," 283; Ludowijck Le Vasseur [Drélincourt], *De Sylviano humore triumvirali epistola* (Paris and The Hague, 1668). On Nuck's letter, "Advies van de Medische faculteit over privaat-colleges van niet-professoren," 22 March 1690, in *Bronnen tot Gescheidenis der Leidesche Universiteit*, ed. P. C. Molhuysen, (The Hague: Martinus Nijhoff, 1920), 4:23*.

[22] Herman Boerhaave, "On Commending the Study of Hippocrates," 74–75 in *Boerhaave's Orations*, eds. E. Kegel-Brinksgreve and A. M. Luyendijk-Elshout (Leiden: Brill, 1983).

[23] Boerhaave, "Study of Hippocrates," 80.

[24] Ibid., 75.

[25] Boerhaave, "On Chemistry Expunging Its Errors," 207 in *Boerhaave's Orations*.

[26] Ibid., 208.

[27] Boerhaave, *Elements*, 53.

[28] Voenno-Meditinski Akademii (hereafter VMA), St. Petersburg, Russia, Fundamental Library, Fund XIII, MS 3, "Collegium Chemicum," fol. 24r: "Effecta separationis ad nullas similes partes, neque ad Elementa revocari queunt, neque sine partium immutatione semper producuntur sola separatione."

[29] Boerhaave, *Elements*, 46.

[30] Ibid.

[31] Ibid.

[32] Boerhaave, *Elements*, 168–201. On Boerhaave's conception of combustion see Love, "Instrument-Element Concept of Fire."

[33] Boerhaave, *Elements*, 170–81.

[34] Ibid., 47, 181–82, and 184–86.

[35] Ibid., 186–91; quotation on 191.

[36] For the "Spiritus rector" experiments, see Boerhaave, *Elements*, 47–49; on Boerhaave's "Mercury" (the alchemical "mercury" principle), see John C. Powers. "Herman Boerhaave and the Pedagogical Reform of Eighteenth-Century Chemistry," Ph.D. Dissertation, Indiana University, 2001, 121–33.

[37] Boerhaave, *Elements*, 47.

[38] For his lecture course, see Boerhaave, "Collegium Chemicum," fols. 1r–93v; for lists of operations presented in Boerhaave's demonstration course, see fols. 143r–78v.

[39] This dichotomy was exemplified by Boerhaave's two medical textbooks, his *Institutiones medicae*, which was based on his theory course in medicine, and his *Aphorisms*, which presented rules of thumb for treatment based on his *praxis medica* course; see Boerhaave, *Institutiones medicae* (Leiden, 1708) and *Aphorismi de cognoscendis et curandis morboris* (Leiden, 1709).

[40] For example, the tenth edition of Nicolas Lemery's extremely popular *Cours de chymie* devoted 68 pages to definitions, principles, and apparatus and 860 pages to operations, Nicolas Lemery, *Cours de chymie*, 10th ed. (Paris, 1713). On the problems with didactic chemistry in Leiden more generally, see John C. Powers, "Chemistry Enters the University: Herman Boerhaave and the Reform of the Chemical Arts," *History of Universities* 21, 2006, 77–116.

[41] Johannes Bohn, *Dissertationes chymico-physicae, chemiae finem, instrumenta et operationes frequentiores explicantes* (Leipzig, 1685). Note that the dates of the exercises and name of each student *defendens* were printed at the beginning of each set of theses.

[42] On Bohn, see K. Rothschuch, "Bohn, Johannes," in *Dictionary of Scientific Biography*, ed. Charles Gillispie (New York: Scribners, 1973); Francesco Trevisani, "'Ratio' und 'Experimentum:' Johannes Bohn (1640–1718) und die Italienische Experimentale Physiologie," *Clio Medica* 17, 1983, 199–206.

[43] Bohn, *Dissertationes*, "De Corporum Dissolutione," paragraphs 12–17. Note that each dissertation is unpaginated.

[44] Bohn, *Dissertationes*, "De Ignis," paragraph 1.

[45] "Ad chymico-physicas Bohn Observationes" in Boerhaave, "Collegium Chemicum," fols. 94r–117r.

[46] Bohn, "De Ignis," paragraphs 19–21. On Mayow's "aerial-niter," see Henry Guerlac, "John Mayow and the Aerial Nitre, Studies in the Chemistry of John Mayow – I," 332–49 in *Actes du Septième Congrès International d'Histoire des Sciences* (Jerusalem, 1953); Robert G. Frank, Jr., *Harvey and the Oxford Physiologists: A Study of Scientific Ideas and Social Interaction* (Berkeley, CA: University of California Press, 1980), 224–74.

[47] Boerhaave argued that the pressure of the air kept the particles of combustibles in close proximity to one another, allowing fire (the instrument) to work effectively. See Boerhaave, *Elements*, 185 and 205–07.

[48] Boerhaave, "Collegium chemicum," fols. 1r–93v.

[49] Walter Ong, *Ramus, Method and the Decay of Dialogue* (Cambridge, MA: Harvard University Press, 1958); Neal W. Gilbert, *Renaissance Concepts of Method* (New York: Columbia University Press, 1960); Peter Dear, "Method and the Study of Nature," 147–77, on 147–50 in *The Cambridge History of Seventeenth-Century Philosophy*, eds. Daniel Garber and Michael Ayers (Cambridge: Cambridge University Press, 1998).

[50] Boerhaave, "Collegium chemicum," fol 28v.

[51] These lectures are recorded in Boerhaave, "Praelectiones chemiae," VMA, Fund XIII, MS 7.

[52] Powers, *Herman Boerhaave*, 162–206.

[53] Boerhaave, "Praelectiones chemiae," fols. 3r–5r, and *Elements*, 79–103.

[54] Boerhaave, "Praelectiones chemiae," fols. 15v–18r, and *Elements*, 214–22.

[55] Boerhaave, *Elements*, 222.

[56] Boerhaave, "Praelectiones chemiae," fol 31r.

[57] Boerhaave, *Elements*, 308–15, quotations on 314.

[58] Ibid., 291.

RINA KNOEFF

PRACTICING CHEMISTRY "AFTER THE HIPPOCRATICAL MANNER"

Hippocrates and the Importance of Chemistry for Boerhaave's Medicine

It is well known that Herman Boerhaave, the eighteenth-century "instructor of all of Europe" (*communis Europae praeceptor*), was an ardent supporter of Hippocrates. While historians have discussed Boerhaave's veneration for Hippocrates as the "Father of Medicine" before, it is less known that Boerhaave also recommended practicing chemistry "after the Hippocratical manner." Boerhaave's advice is remarkable since chemistry is alien to the Hippocratic writings. What, I ask in this paper, did Boerhaave mean when speaking about the "Hippocratical manner" and why did he make this method central to his chemistry? What in the Hippocratic corpus was of particular use in the chemical laboratory that attracted Boerhaave? How did Hippocrates function as an essential connection between Boerhaave's chemistry and medicine?

Boerhaave is known as the systematizer of medicine and chemistry – that is, Boerhaave transformed them into (teaching) disciplines with clear methods, aims, materials, procedures, topics, questions and answers.[1] In so doing Boerhaave initiated a new, perhaps even revolutionary, method of teaching and research that, via his students, spread all over Europe. In the history of chemistry, however, Boerhaave's novel approach has gone largely unnoticed. As Seymour Mauskopf and Lawrence Principe have argued elsewhere in this volume, the main focus of attention has been the Chemical Revolution centered around the career and successes of Lavoisier. Not only has early eighteenth-century chemistry been ignored, but, as Principe convincingly shows in the case of the French chemist Homberg, it has been rewritten in order to suppress the sort of chemistry and alchemy that was later considered unscientific and even fraudulent. Much of Boerhaave's chemistry (and particularly those parts connected to alchemy and religion) was likewise ignored or considered unfortunate slips of the pen.[2] Only recently has Boerhaave scholarship recovered the arcaner part of Boerhaave's work and so shed a new light onto the work of the Dutch reformer of chemistry and medicine.[3]

In Boerhaave's transformation of chemistry into an accepted academic discipline "Hippocrates" played a key role. Not only did the Hippocratic Corpus shape Boerhaave's chemistry for medicine, but for Boerhaave, Hippocrates was a perfect role model for chemistry as a discipline. John Powers has argued that Boerhaave chose Hippocrates because in this way he could make chemistry acceptable to a medical faculty oriented towards Hippocratic teaching.[4] However this may be, I argue that there was more to it than just political and pragmatic reasons. Like Principe, I maintain that

L. M. Principe (ed.), New Narratives in Eighteenth-Century Chemistry: Contributions from the First Francis Bacon Workshop, 21–23 April 2005, 63–76.

early modern chemistry was often informed by implicit ideas, natural to the chemist – and therefore not explicitly mentioned – but alien to the modern historian. In the case of Boerhaave, religion was such an implicit factor influencing not only his way of life, but also the way he thought about nature and natural philosophy. I argue that the reason Boerhaave chose Hippocrates as role model in chemistry, was not only because he was a handy leg up to the medical faculty, but also because more than any other natural philosopher Hippocrates answered Boerhaave's Calvinist idea of natural philosophy. Hippocrates, according to Boerhaave, viewed nature as providential and he pleaded for meticulous observation. Furthermore the supposed Hippocratic meaning of "Nature" is strikingly similar to Boerhaave's Calvinist understanding of the category "Nature." Moreover, the Hippocratic image of human life invigorated and sustained by vital forces, inspired Boerhaave to make chemistry basic to his medicine.

BOERHAAVE'S CALVINIST HIPPOCRATES

From the beginning of his medical studies in 1691, Boerhaave was convinced of the importance of the Hippocratic corpus for medicine.[5] In the *Commentariolus*, the autobiographical notes found after his death, he wrote:

> He [Boerhaave] began his reading of the ancient medical writers in chronological order, starting with Hippocrates; soon he understood that the later authors owed to Hippocrates everything that was good in their work; therefore to him alone he devoted a long time, reading him, summarizing him. Running through the more recent writers he halted at Sydenham, whom he worked through several times, each time more eagerly.[6]

Throughout Boerhaave's career the Hippocratic corpus remained at the center of his teaching. Of his ten orations the first and the last were specifically devoted to Hippocrates, while even in the other orations he often alluded to the works of the "Father of Medicine." Hippocrates was so central to Boerhaave's medical system that in his view "before a physician can be of any real use to the sick, he should have studied these works [of Hippocrates, and] his hands should have leafed through them night and day."[7] Not surprisingly, Boerhaave's medical works are full of references to Hippocrates. Even more, Boerhaave *identified* with the ancient author, copying the layout of the Hippocratic corpus. Most notably Boerhaave's *Aphorismi de cognoscendis et curandis morbis* are reminiscent of the other (Hippocratic) aphorisms that medical students were still asked to explain during their examinations. Not only did Boerhaave model his medical system on what he considered to be genuinely Hippocratic, his followers also answered in a "Hippocratic" manner. Most notably the commentaries of Albrecht von Haller and Gerard van Swieten upon Boerhaave's *Institutiones medicae* are reminiscent of a scholastic manner of explaining the Hippocratic writings. In large print they quote Boerhaave's statements, following each with a much longer exposé in small print. Thus, Boerhaave not only developed a medical system according to lines he considered Hippocratic, but also to his followers Boerhaave himself became a new Hippocrates.[8]

Wesley Smith has already acknowledged the decisive influence of Boerhaave on the modern image of Hippocrates. He argued that Boerhaave "who had such an effect on ideas of medicine, also affected medical history," and he did so in such a way that "his outlook became *the* view, as though there was no other way in which the past could be conceived."[9] Andrew Cunningham has similarly argued that

> ... what Boerhaave inevitably did was to *take* to Hippocrates, and other writers, certain attitudes from his own day. As a result he makes of both ancients and moderns some quite new persons ... Boerhaave's success in creating these new images for the dead was so great that we today tend to hold Boerhaavian views of all the ancients and moderns of whom Boerhaave approved.[10]

Smith's and Cunningham's conclusions fit recent scholarship on the Hippocratic tradition, which starts from the assumption that "Hippocrates is not so much a 'real' person, as a malleable cultural artefact, constantly moulded and remoulded according to need."[11] It follows that, in order to understand the Boerhaavian Hippocrates, we have to study Boerhaave's motives for adopting Hippocrates as his main hero. I argue that these motives were Calvinist and that Boerhaave saw Calvinist ideals paradigmatically expounded in Hippocrates.[12]

The Calvinism of Boerhaave's natural philosophy is visible in two ways.[13] First, Boerhaave aimed at showing the wisdom of God in the divine works of creation and providence. For instance, Boerhaave's image of pure fire as a divine instrument through which God steers His creation is based upon a radical belief in divine providence. In Calvinist terminology, Boerhaave believed that God presides over Nature, he holds the helm and rules over absolutely all events. This means that natural philosophers have a privileged insight into God's working hand in nature and in His divine plan with the world. Secondly, Boerhaave time and again contrasted the omnipotence of God with the littleness of man and he emphasized the limitations of the human intellect and investigation on every occasion. As a direct result, Boerhaave rejected the possibility of capturing the working of nature in a few natural laws (as Descartes and Newton had done), and he devoted his research to the observation of the smallest and most peculiar powers of natural bodies. The individuality of natural phenomena made him realize that trying to understand the works of God via the method of mathematical reason was not only too ambitious, but also entirely impossible. Boerhaave considered God's divine works too great and too complex to fit into the enclosure of the human mind.[14] Moreover, Boerhaave's negative view of the capacities of the human intellect made him favor the method of observation, for only in this way true knowledge of the outside world could be imprinted on the human mind.

Particularly Boerhaave's insistence on keeping to the study of the *individual* characteristics and powers of nature set him apart from other religiously inspired natural philosophers. It can be argued that the idea of God's handiwork manifested in nature was a regular feature of all natural theology at the time and also that the emphasis on the weakness of man's intellect can be found in the works of other, non-Calvinist, natural philosophers. Yet, Boerhaave's aversion to general theory, resulting from these two premises, is uniquely Dutch Calvinist. While, for instance, English Puritan natural philosophers tried to unveil the hidden essence of nature in order to restore paradise on earth, Boerhaave believed it impossible to ever recover the first principles of nature.[15] Unlike for instance Newton, Boerhaave never believed he could proceed to the First Cause of things and be able to answer questions about the design and purpose of the universe. In 1715, after handing over his office as *Rector Magnificus* of the University, Boerhaave stated:

> Yet I would like to ask anyone whether he understands all this? Whether he comprehends in his mind the immense thing that is the universe? ... We would be ashamed to find anyone so lacking in self-knowledge, so destitute of modesty, as to take this upon himself. For then he would in monstrous stupidity be aiming at the wisdom itself of God, and equal the giants in arrogant pride.[16]

Boerhaave's emphasis on the peculiar characteristics of bodies as a means to avoid general theory was shared by his Dutch contemporaries. Harold Cook has argued that Dutch natural philosophy was not directed towards constructing a mechanical worldview, but can be characterized as the exploration of detail.[17] Edward Ruestow has likewise argued that the microscope had an extensive influence on Dutch cultural life since it "testified to an evermore intricate complexity in nature and a pervasive and continued unexpectedness."[18] Moreover, the Dutch arts likewise reveal a great attention to detail. This feature, according to Svetlana Alpers, resulted from the Dutch visual rather than textual culture that in her view also explains the Dutch attention to experiment and observation.[19] With the exception of Ruestow, no one has connected the Dutch preoccupation with detail to Calvinist motivations. However, Boerhaave's reasoning shows that there was a direct connection between the Calvinist emphasis on a God who is in command of absolutely everything on earth, the dependence of man on his Creator, and man's limited capacity to research only the smallest powers of nature – for these should suffice to offer man a glimpse of the divine wisdom without the danger of pretending to know God Himself.

With these Calvinist premises in mind, Boerhaave read and interpreted the Hippocratic corpus. As already suggested by Andrew Cunningham, Boerhaave considered the works of Hippocrates as the original uncorrupted source of medicine much like the Bible was for Protestant theology.[20] It is therefore not surprising that for Boerhaave Hippocrates already incorporated the essential characteristics of good (read Calvinist) research. What he praised most in Hippocrates were the two essential Calvinist characteristics mentioned above: the centrality of teleology in Nature – that in Boerhaave's view equals God's providential creation – and the meticulous observation of detail.

Nowhere is Boerhaave's Calvinist Hippocrates more visible than in his last oration on servitude as the physician's glory (1731). After introducing the theme of his speech, the physician as a humble servant of nature, Boerhaave went on defining Nature as God's creation and he explains God's command over all living creatures:

> All things that come within the range of human thought relate either to God or to the physical Universe. We venerate the Highest Power as being eternal, one

> and only, the mainspring of all other phenomena. It has been granted to mortal beings to adore and love His Majesty; beyond this nothing more is possible for us, because He is only known to Himself. The Universe of Nature, however, created by Jove [= God] in a manner that is both incomprehensible and ineffable, contains things, each of which has been endowed with its singular structure; through this they are individually defined ... if we now look more closely at this earth of ours, a true planet, we learn that it is composed of living beings, plants and minerals; fire, air and water; and that all this is again subject to the firm rule and unassailable maxims of the Creator of Nature; everything obeys His commands.[21]

In the same oration Boerhaave also argued that Hippocrates was among the first to acknowledge God's divine providence in nature. When speaking about the essentially incomprehensible nature and movement of the bodily fluids, Boerhaave argued that

> This is something beyond our most skilful efforts. Hippocrates pondered all this carefully and then exclaimed in amazement: "They know not what they [the humours] are doing, but they appear to know it!" Because they come into being through divine disposition they execute the design, as established by Jove [= God].[22]

Having established that God's omnipotent and incomprehensible steering hand is active in nature, Boerhaave states that man is unable to fundamentally understand "anything at all about even the minutest particles of the ingenious structure of the body." The natural philosopher is only able to know what is granted to him by nature herself – and nature can only reveal herself by means of sense-perception.[23] Thus only through careful observation is man able to gain (partial) knowledge of nature. Again it was Hippocrates who, according to Boerhaave, in good Calvinist manner, was devoted to observation, for "he never invented what he had not actually observed: that he never failed to note what there was to observe; that he never twisted or meddled with the truth, when describing the works of nature."[24]

Boerhaave was not the first to portray Hippocrates as the founder of the modern observational method. Particularly in the seventeenth century, Hippocrates became a symbol of empirical research as opposed to Galenic rationalism and theory. David Cantor has argued that the very nature of the Hippocratic Corpus enabled a shift from Galen to Hippocrates. Since the Corpus was written by different people in different times, it was possible to change the Galenic set of Hippocratic writings for another, thereby abandoning Galen, while keeping Hippocrates. Moreover, seventeenth-century physicians were increasingly interested in rules of conduct and methods in medicine. Hippocrates, in their eyes, offered timeless classical values.[25] The influential Italian doctor Giorgio Baglivi (1668–1707) is a good example of how Hippocrates was remodelled according to seventeenth-century needs. In order to avoid the disputes over medical theory and ancient authority he proposed to turn to medical practice. His work was directed to the formulation of "modern" aphorisms based on careful descriptions and interpretations of histories of diseases. Baglivi's aphorisms resemble those of Hippocrates, but include causes, which the Hippocratic aphorisms do not. So Baglivi included natural philosophy, of major importance in seventeenth-century research, into his medical system while at the same time holding on to Hippocrates as his primary role model.[26]

Boerhaave read Hippocrates in a similar way. Although the ancient Greeks had no exact equivalent for our word "observation," for Boerhaave Hippocrates was the first and foremost advocate of this very method in natural philosophy.[27] In good Calvinist manner, Boerhaave linked Hippocrates' method of observation to the greatness of the divine creation as opposed to the littleness of man and his inability to come to true knowledge via the method of (Cartesian) reasoning. Hippocrates, in Boerhaave's view, united all the good (Calvinist) qualities that he ideally wanted to see in his students. Hippocrates "never invented what he had not actually observed"; he used "a simple style which sets forth the subject matter briefly and clearly"; and he never lost himself in "dubious fantasies about the four elements or about the four primary qualities," but "penetrated the true occurrences in reality."[28] Boerhaave told his chemistry students likewise that experiments "should be simple, or not compounded of various concurrent operations" and that "the changes it makes should not be very remote from the nature of the subject."[29]

Of all natural philosophers, Hippocrates best incorporated the Calvinist way of doing research. And it was to Hippocrates, in Boerhaave's words, that all "the later authors owed everything that was good in their work."[30] Hence Boerhaave recommended that his students practice chemistry "after the Hippocratic manner." This means that Hippocrates was Boerhaave's first and foremost role model in his academic pursuits. So even though Boerhaave presented Francis Bacon as a role model for the natural philosophy of his day (Bacon's experimental method, after all, exactly fitted Boerhaave's Hippocratic model), for Boerhaave it was a Calvinist Hippocrates who, before Bacon, made observation central to medicine, thereby establishing the right method for natural philosophy as a whole.[31]

"NATURE" AND BOERHAAVE'S CHEMISTRY OF LIVING BEINGS

Before discussing Boerhaave's chemical practice "after the Hippocratical manner" and its importance for medicine, it is necessary to look briefly at Boerhaave's understanding of "Nature." This is necessary first and foremost because Boerhaave considered chemistry as the way best suited to observe and investigate "Nature."

According to Luyendijk-Elshout and Kegel-Brinkgreve, in their translation of Boerhaave's orations, Boerhaave's "Nature" had three meanings: first, it stands for the sum total of phenomena in the world; second, Nature is about the regular movements of these phenomena, and the laws that rule them; and third, Nature stands for a vital force that is innate in living beings.[32] Particularly the third aspect concerns us here since Boerhaave's chemistry was mostly concerned with recovering these latent peculiar powers of bodies, i.e. the (vital) forces responsible for motion and change. Again it was Hippocrates who in Boerhaave's view first recognized the importance of these forces. Whether speaking about the *impetum faciens*, "that which imparts," or the *vis medicatrix naturae*, the spontaneous healing power of nature, Hippocrates, according to Boerhaave, acknowledged that "Nature is sufficient to herself in all situations; that she imparts, governs, and maintains the vital functions; that she never undertakes anything foolish or aimless, and favours the normal."[33]

It is in this line of argument that Boerhaave used the metaphor of a circle to define the nature of the body. He argued that

> ... whenever the expert in his scrutiny focuses on one single part of the body, he will always find that he needs to know the other parts as well. For the isolated part that he chose for examination is linked to all the other ones. There is no beginning and no ending here; *the human body is a circle in which beginning, middle, and end are everywhere identical.* No part is created later by another one which came into being previously; they are all born at the same time, intertwined with one another – I grant that – each with its individual form.[34]

Again, Boerhaave almost literally cited the Hippocratic Corpus, where in various places the metaphor of the circle is used to denote the human body. For instance in *Places in Man*, Hippocrates argued that "the body has no beginning, but everything is beginning and end alike: for when a circle has been drawn, its beginning cannot be found."[35] This Hippocratic passage has often been quoted in early modern medical literature in order to show that Hippocrates already knew about the mechanical circulation of the blood and other humors in the body. Most notably Iain Lonie has argued that Hoffmann quoted Hippocrates as an early adherent of iatromechanism.[36]

Yet, the Hippocratic passage goes on in a "vitalist manner" stating that all things in the body "flow" and "breathe" together "in sympathy" – "all parts in the whole, and the parts in each particular part, with a view to the function." This passage, according to Lonie, was most certainly influenced by the Stoic doctrine of cosmic "sympathy," according to which all parts in the cosmos are related to one another as in a living body.[37] Stoic doctrine not only influenced the Hippocratic writings, but it was also influential in Calvinism. Calvin himself was known as a commentator of Seneca, and Stoic ideas were particularly fruitful in Calvinist contexts both in the Scottish Enlightenment as well as in the Netherlands.[38] It is therefore no surprise that we find Calvin stating that God rules the universe by His "divine energy," and that "all the parts of the world are invigorated by the secret inspiration of God."[39] The Calvinist Boerhaave, in turn, similarly argued that God infused active principles in lifeless matter through which he actively steers His creation as an organic whole. With respect to the body, Boerhaave most clearly stated that God is directly and solely in charge of its working. He argued that

> ... o, mortals, you should worship God, who has ordered these solid and fluid substances in one structure in such a manner that through its unique powers He can eventually replace lost parts by wholly similar ones; but He has made it wholly impossible for any other cause to achieve the same result.[40]

It is no surprise that Boerhaave was attracted to Hippocrates, who, in his view, presented a view of nature completely in accordance with his own ideas on divine creation and providence. Indeed, Boerhaave's adoption of the Hippocratic image of the body as a circle has a religious undertone, for Boerhaave continued his argument stating that:

> ... even the tiniest part coming into being requires that the body in its entirety should have been created beforehand .... Indeed, it is contrary to divine law that even the sum-total of human science should be able to bring forth one hair from a healthy body if there were not a multitude of vessels, instruments, and humours at hand for this purpose, which had been created by Nature's Maker at the very beginning, in order that they might bring forth this same individual hair.[41]

As Hippocrates, and in line with the argument that the body is like a circle, Boerhaave maintained that all substances in the body are essentially the same and mutually dependent: "nothing is redundant or accidental."[42] In almost all of his medical works Boerhaave started his account of the body with an explanation of its fibers. He described them as long threads built of earth-like particles without a hollow space. The fibers are woven together into membranes in the same way that threads of linen are woven together into a cloth. When you roll the membrane, small vessels appear (*vasa minora*), and these small vessels can again be woven together into bigger vessels (*vasa maiora*), which in turn build up the organs and other parts of the body. This means that all the parts of the body, even the bones, consist of vessels. The smallest vessels, however, are invisible and their existence can only be deduced.[43] All the vessels are directly or indirectly linked to the left ventricle of the heart, so that all fluids (with the exception of excretions) return to the heart. As long as the fluids move regularly through the channels, the body is alive. As soon as the movements are obstructed or cease altogether, the body falls ill and ultimately dies.

So far, Boerhaave's description of the body is rather mechanical. He described the body as a *perpetuum mobile*. Life is closely connected to motion, and motion depends on the laws of mechanics, hydrostatics, and hydraulics. Yet Boerhaave argued at the same time that the body is an organic unit steered by vital forces inherent in its physical substances. Therefore he stated in his lectures that not mechanics, but chemistry would be of the utmost importance in understanding the nature of the body: "if chemistry did not exist it would be impossible ... to gain insight into the proper nature and forces of single bodies."[44] This means that according to Boerhaave, chemistry is better suited to investigate the inner workings of nature. For this reason he made chemistry a central discipline in medicine.

Once again it was Hippocrates who inspired Boerhaave to do so, for Hippocrates first realized that in order to cure the sick, the physician should have a thorough understanding of the structure of the body, its simple and composite, harmful and salubrious parts. Only then he is "sufficiently knowledgeable to record the useful qualities of vegetable, mineral, and animal substances." Boerhaave argued that this is "the only way to assist the sick ... the way trod by the founder of our science, to which he guided us by word and deed."[45] Boerhaave knew that Hippocrates did not mention chemistry as a means to knowledge about the body, but he argued that physicians in the tradition of Hippocrates did. For instance, he mentioned Van Helmont as a true follower of Hippocrates because Van Helmont, in the footsteps of Hippocrates, employed the observation of "Nature" as a central concept in his natural philosophy.[46]

Boerhaave was particularly interested in Van Helmont's application of chemistry to the investigation of the nature of the bodily fluids, "which are responsible for the functions of motion and life in us."[47] In other words, Boerhaave constructed his argument so that he made Hippocrates an early supporter of the chemical method in medicine *avant la lettre*.

Most importantly, Boerhaave analyzed the fluids in a chemical way, that is, he separated them into their smallest parts and qualified the parts as sharp, acid, alkaline, salty, and so forth. The particular qualities of the particles are closely connected to powers of cohesion, attraction, fermentation and so forth that cause chemical reactions between the particles. These chemical reactions in turn are of crucial importance for the motion of the humors inside the solid vessels. Illnesses are most often caused by a change in the nature of the particles. For instance, Boerhaave believed that many illnesses are caused by putrefaction of the blood stagnating in an obstructed vessel. The otherwise perfectly round and neutral particles change into an acid or alkaline nature and harm the body.

Boerhaave, however, not only analyzed fluids in a chemical way, but also explained bodily processes at large in terms of chemistry. The process of digestion is not only an obvious example, but was also central to Boerhaave's medicine. He started his physiology in the *Institutiones medicae* with the process of digestion, for he believed that the body is made of the food it ingests.[48] The logic of starting with the process of digestion (which in chemical terms equals the process of changing food and drink into body) is even visible in the second part of Boerhaave's chemical textbook, the *Elementa chemiae* which is concerned with the operations of chemistry. He maintained that all animal substances (flesh, blood, milk and ultimately butter, cheese and meat) are built of vegetable matter (such as grass and hay) through the process of digestion, so it is logical to speak about the vegetable kingdom before discussing animal bodies.

At the time it was not unusual to explain the process of digestion in terms of a reaction between acid and alkaline particles. Boerhaave, however, refuted this thesis. On several occasions he showed that the juices of a healthy animal (saliva, stomach liquor, bile, pancreatic juice, milk, urine, egg white and blood serum) do not show any signs of effervescence when mixed with acids or alkalis. Boerhaave argued instead that the bodily fluids are neutral and therefore that the chemical process of fermentation in the digestive tracts is directed towards the neutralization of the body's acid and alkaline tendencies. Experiments on urine, for instance, had shown that the bodily particles have a spherical figure so as not to damage any tissue. To illustrate his argument Boerhaave pointed to the fact that the urine of a man who drank and ate large quantities of Rhenish beer, vinegar, and fruit did not show any sign of acidity.

Boerhaave's description of the process of digestion shows that he did not consider the transformation of food into bodily materials a purely mechanical process. Instead, he based his account on the way particles react when they are brought into contact with one another. Even though Boerhaave condemned iatrochemists who immediately applied the results of chemical experiments to the working of the body, he himself had no problems applying *in vitro* experiments to *in vivo* situations.[49] He did so mainly through describing his chemical experiments as well as the processes of the body in terms of the forces of cohesion, attraction, fermentation, etc. Moreover, he did not solely consider these forces of motion in a mechanical way since he argued in a (late) Newtonian manner that the transformation of bodies by means of motion is "the subject of chemistry, and of chemistry alone."[50] In so doing Boerhaave clearly distinguished between the disciplines of mechanics (which in a Newtonian sense are synonymous with physics) and chemistry, a distinction that had been matter of debate from the seventeenth century onwards. In the 1960s the distinction was that "chemists love molecules, and get to know them individually, in the same way that politicians love people." In contrast ... "physicists are more concerned with fields of force and waves than with the individual personalities of the molecules or matter."[51]

This distinction is a modern one – we cannot speak about molecules as such when speaking about the eighteenth century – yet it is still a useful definition for clarifying Boerhaave's understanding of chemistry and mechanics. Boerhaave likewise argued that chemistry is concerned with the individual powers of particles of matter, while mechanics or physics is more concerned with general theory. In the *Elementa chemiae*, Boerhaave maintained that

> Mechanics, and those skilled in hydrostatics and hydraulics, have explained many operations in nature by an infallible method, from the general properties common to all bodies. But from all these sciences, how much soever improved, they have never been able to account for those effects of bodies which depend on the disposition peculiar to certain kinds thereof; which the Creator has endowed therewith beyond all the rest; as those effects would never have existed, had such peculiar power of property of the body been wanting.[52]

In accordance with this definition, when considering the body from the angle of chemistry, Boerhaave was primarily interested in the individual powers through which compounds are resolved into simples. This means that, for instance, in the case of the blood circulation, he was not so much interested in the mechanical (machine-like) motion of the blood through the veins and arteries, but was much more concerned with the question of how the particles of arterial blood break up and change into venous blood, nervous liquid, and lymph. It also means that Boerhaave did not put the description of general mechanical laws of nature high on his agenda. Instead, he believed that only through understanding the *individual* powers of bodies, as studied by chemists, can the physician obtain better insight into the workings of nature. In other words and in line with his Calvinist beliefs, Boerhaave considered the chemist's method of studying particulars most suitable in medicine.

The emphasis on the individuality of natural phenomena is particularly evident in Boerhaave's educational programme. Rather than presenting his students ready-made recipes and formulae, he stressed time and again that physicians in the tradition of Hippocrates have to study the first rudiments and methods of a particular discipline. In chemistry Boerhaave emphasized that it is much more important to understand what chemistry is and does, instead of blindly following chemical formulae. He encouraged his students to carefully observe and contemplate the nature of things before preparing a remedy or setting up a chemical experiment.[53] Moreover, in his oration on the simplicity of purified medicine, Boerhaave argued that physicians following Hippocrates and Sydenham ("the English Hippocrates") have shown that the application of a cure ultimately depends upon a thorough understanding of the nature of the body and its diseases. In *Nature of Man*, Hippocrates argued that "disease has a plurality of forms and a plurality of cures." Boerhaave in turn stated that Hippocrates "did not bother to compare the variety [of remedies] – which is useless – but [wanted] to get to know when they were really necessary and opportune – which is the heart of the matter – and pondered what was necessary and in what way it must be administered."[54] Additionally, Boerhaave repeated after Sydenham that "he who knows *what* to do, is seldom at a loss as to the required remedy."[55] Hence, it is not surprising that Boerhaave's medical works have the same style as those of Hippocrates – they do not prescribe remedies for particular diseases, but they consist of statements of wisdom, such as "Desperate cases need the most desperate remedies" (Hippocrates), or "the most simple diseases can be traced back to the most simple fibers" (Boerhaave).[56]

CONCLUSION

Instead of maintaining that Boerhaave's medical system was a bit of both mechanics and chemistry, as historians have done in the past, I make the much stronger claim that chemistry was basic to Boerhaave's medical teaching.[57] That is not to say that Boerhaave wanted to explain medicine in terms of chemistry alone like the iatrochemists had done before, or that mechanical disciplines were of no concern. I merely quote Boerhaave's words in stating that

> ... chemistry surpasses all disciplines in usefulness [after it has been purged of its errors of the past]. In physics we can be of good cheer with this guide, in medicine all possible good may be expected from it. It teaches most faithfully how the deepest secrets may be revealed, intricacies be disentangled, how hidden forces of bodies may be discovered, imitated, directed, changed, applied and perfected.[58]

In 1718 Boerhaave argued that "anybody who now trains his mind to follow the precepts of this discipline [chemistry] ends up by having a refined insight into the secrets of nature and medicine."[59] After all, chemistry offered the right methods and tools to break up and change natural bodies, thereby disclosing the individual powers through which each body lives and moves.

Moreover I have shown that Hippocrates functioned as a role model for Boerhaave's chemistry. According to Boerhaave, the Hippocratic Corpus ideally reflected his Calvinist ideas on Nature in general and on the human body in particular. No one better than Hippocrates represented the right understanding of divine and providential Nature, and no one better understood the importance of keeping to the study of particulars. Boerhaave's Hippocratic method in medicine and chemistry was influential; university courses across Europe were modelled on Boerhaave's method. In spite of the historiographic emphasis on *scientific* achievements leading up to Lavoisier, in

Boerhaave's case it was *religious* motivations that inspired a new method of chemistry, which in its time was at least as important as the so-called Chemical Revolution launched by Lavoisier.

NOTES

1 For an evaluation of the early modern meaning of "discipline" see Andrew Cunningham, "The Pen and the Sword: Recovering the Disciplinary Identity of Physiology and Anatomy before 1800. I. Old Physiology – The Pen," *Studies in History and Philosophy of Biomedical Sciences* 33, 2002, 631–65, on 632–33.

2 See for instance Gerrit A. Lindeboom, *Herman Boerhaave: The Man and His Work* (Leiden: Brill, 1968), Boerhaave's chief twentieth-century biographer. But also more recently historians have denied that Boerhaave's natural philosophy included immaterial powers; see for instance Harold J. Cook, "Boerhaave and the Flight from Reason in Medicine," *Bulletin for the History of Medicine* 74, 2000, 221–40.

3 Andrew Cunningham, "Medicine to Calm the Mind. Boerhaave's Medical System and Why It Was Adopted in Edinburgh," 40–66 in *The Medical Enlightenment of the Eighteenth Century*, Andrew Cunningham and Roger French, eds. (Cambridge: Cambridge University Press, 1990); John C. Powers, *Herman Boerhaave and the Pedagogical Reform of Eighteenth-Century Chemistry* (Ph.D. dissertation; Indiana University, 2001); Rina Knoeff, *Herman Boerhaave (1668–1738). Calvinist Chemist and Physician* (Amsterdam: Edita, 2002).

4 See John Powers's contribution to this volume.

5 The exact date of the beginning of Boerhaave's medical studies is unknown. Lindeboom (*Herman Boerhaave*, 28) has suggested that it was in 1691 because Boerhaave, after graduating with a philosophy degree in 1690, worked in the university library for nine months.

6 Herman Boerhaave, *Commentariolus*, XII; translated in Lindeboom, *Herman Boerhaave*, 381.

7 Herman Boerhaave, *Oratio de commendando studio Hippocratico* (Leiden: 1701), 6; translated in A.M. Luyendijk-Elshout and E. Kegel-Brinkgreve, *Boerhaave's Orations* (Leiden: Brill, 1983), 66. In the following I shall refer to the translation of the oration as CSH.

8 Cornelis Love, translator of Boerhaave's *Aphorisms*, explicitly praised Boerhaave as the "Dutch Hippocrates." See his preface to *Kortbondige Spreuken wegens de Ziektens te Kennen en te Genezen* (Amsterdam, 1741). Recently, Jan K. van der Korst noted that perhaps Boerhaave rather than Pieter van Foreest earned the title of "the Dutch Hippocrates"; "Van gevierd Hippocraat tot vergeten Galenist," 145–52 in *Pieter van Foreest: De Hollandse Hippocrates*, Henriette Bosman-Jelgersma, ed. (Krommenie: Drukkerij Knijnenberg,1996), on 145.

9 Wesley D. Smith, *The Hippocratic Tradition* (Ithaca, NY: Cornell University Press, 1979), 27.

10 Cunningham, "Medicine to Calm the Mind," 49. Elsewhere Cunningham argues that it was Sydenham's representation of Hippocrates that Boerhaave spread across Europe. I disagree with this interpretation. I argue in this paper that Boerhaave's Hippocrates was Calvinist and that what he adopted from Sydenham's understanding of Hippocrates primarily fitted the Calvinist criteria of his research. See Andrew Cunningham, "The Transformation of Hippocrates in Seventeenth-century Britain," 91–115, on 105 in *Reinventing Hippocrates*, David Cantor, ed. (Aldershot: Ashgate, 2002).

11 David Cantor, "The Uses and Meanings of Hippocrates," 1–18 in *Reinventing Hippocrates*, on 3.

12 Other (non-Calvinist) natural philosophers also adopted Hippocrates as their primary role model. Thomas Rütten has argued that Hippocrates was constructed a *Christus medicus* in the medical tradition which developed analogously to the theological tradition, "Hippocrates and the Construction of 'Progress' in Sixteenth- and Seventeenth-Century Medicine," 37–58 in *Reinventing Hippocrates*, on 43. According to Owsei Temkin, *Hippocrates in a World of Pagans and Christians* (Baltimore: Johns Hopkins University Press, 1991), the ease of adopting Hippocrates into a Christian tradition has to do with the Hippocratic suggestion that nature is full of divine powers and that the physician is a servant of this divine nature.

13 In my Ph.D. dissertation, Knoeff, *Herman Boerhaave (1668–1738)*, I have already extensively discussed Boerhaave's Calvinism. See also Knoeff, "The Making of a Calvinist Chemist: Herman Boerhaave, God, Fire and Truth," *Ambix* 48, 2001, 102–11.

14 Knoeff, *Herman Boerhaave*, 6–7.

15 For the English Puritans see Robert K. Merton, *Science, Technology and Society in Seventeenth Century England* (New York: Harper & Row, 1970); Charles Webster, *The Great Instauration. Science, Medicine and Reform 1626–1660* (London: Duckworth, 1975).

16 Herman Boerhaave, *Sermo academicus de comparando certo in physicis* (Leiden, 1715), 6; translated in Luyendijk-Elshout and Kegel-Brinkgreve, *Boerhaave's Orations*, 157.

17 Harold J. Cook, "The New Philosophy in the Low Countries," 115–49 in *The Scientific Revolution in National Context*, Roy Porter and Mikulás Teich, eds. (Cambridge: Cambridge University Press, 1992), on 116.

18 Edward G. Ruestow, *The Microscope in the Dutch Republic: The Shaping of a Discovery* (Cambridge: Cambridge University Press, 1996), 4.

19 Svetlana Alpers, *The Art of Describing: Dutch Art in the Seventeenth Century* (Chicago: University of Chicago Press, 1983), xvii.

20 Cunningham, "Medicine to Calm the Mind," 48.

21 Herman Boerhaave, *Sermo academicus de honore medici, servitute* (Leiden, 1731), 6–7; translated in Luyendijk-Elshout and Kegel-Brinkgreve, *Boerhaave's Orations*, 247–48. In the following I shall refer to the translation of the oration as HMS.

22 Ibid., 252. Luyendijk-Elshout and Kegel-Brinkgreve have searched for the source of the Hippocratic quotation, but they have found none.

23 Ibid.

24 CSH, 70.

25 Cantor, *Reinventing Hippocrates*, 6.

26 Roger French, *Medicine before Science: The Business of Medicine from the Middle Ages to the Enlightenment* (Cambridge: Cambridge University Press, 2003), 207–12.

27 For a discussion of the Greek meaning of "observation" see Geoffrey E. R. Lloyd, *Magic, Reason and Experience: Studies in the Origin and Development of Greek Science* (Cambridge: Cambridge University Press, 1979), 129. See also Cantor's discussion of Lloyd's argument in his introduction to *Reinventing Hippocrates*, 3–4.

28 CSH, 69–71.

29 Herman Boerhaave, *A New Method of Chemistry*, trans. by Peter Shaw, 2nd ed., 2 vols. (London, 1741), 2:3.

30 Boerhaave, *Commentariolus*, XII. The *Commentariolus* is published in Lindeboom, *Herman Boerhaave*, 377–86.

31 For Boerhaave considered as a Baconian see Luyendijk-Elshout and Kegel-Brinkgreve, *Boerhaave's Orations*, 16. More recently Ursula Klein has also argued for a Baconian Boerhaave in "Experimental History and Herman Boerhaave's Chemistry of Plants," *Studies in the History and Philosophy of Biological and Biomedical Sciences* 34, 2003, 533–67, on 557.

32 Luyendijk-Elshout and Kegel-Brinkgreve, *Boerhaave's Orations*, 238.

33 HMS, 261.

34 Ibid., 251, my italics.

35 Hippocrates quoted in Iain M. Lonie, "Hippocrates the Iatromechanist," *Medical History* 25, 1981, 113–50, on 138.

36 Ibid., 138–39.

37 Ibid., 140.

38 Nicholas G. Round, "Alonso de Cartagena and John Calvin as interpreters of Seneca's *De clementia*," 67–88 in *Atoms, Pneuma and Tranquility: Epicurean and Stoic Themes in European Thought*, Margaret Osler, ed. (Cambridge: Cambridge University Press, 1991). See also in the same volume: M. A. Stewart, "The Stoic Legacy in the Early Scottish Enlightenment," 273–96.

39 John Calvin, *Institutes* (Geneva, 1559), I, ix, 4.

40 HMS, 253.

41 Ibid., 251.

42 For Hippocrates see *Nature of Man*. For Boerhaave see HMS, 253.

[43] Herman Boerhaave, *Verhandeling over de Kragten der Geneesmiddelen* (Rotterdam, 1756), 19–44.

[44] Herman Boerhaave, *Sermo academicus de chemia suos errores expurgante* (Leiden, 1718), 40; translated in Luyendijk-Elshout and Kegel-Brinkgreve, *Boerhaave's Orations*, 212. In the following I refer to the translation of this oration as CSEE. Boerhaave similarly argued in the preface to Bellini's *De urinibus et pulsibus* (Leiden, 1730) that the mechanical philosophy deserves praise, but that chemistry is best able to investigate the nature of the humors.

[45] CSH, 74–75.

[46] William Burton, one of Boerhaave's first biographers, recognized the importance of Van Helmont for Boerhaave's work. He stated that Boerhaave "had read over carefully Paracelsus four, and Van Helmont seven times: the latter was his favourite." William Burton, *An Account of the Life and Writings of Herman Boerhaave* (London, 1743).

[47] CSH, 83. Boerhaave was not the first to enlist Van Helmont among the admirers and followers of Hippocrates. Jole Shackelford ("The Chemical Hippocrates: Paracelsian and Hippocratic Theory in Petrus Severinus' Medical Philosophy," 59–88, in *Reinventing Hippocrates*, on 60) has argued that Hippocrates' reputation among the English chemical physicians was presumably partly shaped by Van Helmont's vision of Hippocrates as a chemist.

[48] Herman Boerhaave, *De Geneeskundige Onderwijzingen*, C. Love, trans. (Amsterdam, 1745), 13.

[49] For Boerhaave condemning the iatrochemists see CSEE, 205–07.

[50] CSEE, 199–200. See also Isaac Newton, *Opticks* (London, 1730), Query 31, 375–96, on 394. This is not to say that Boerhaave was a Newtonian *pur sang*. Only in the beginning of his academic career can Boerhaave be called a Newtonian. Later in life he became more critical of the English natural philosopher.

[51] Mary Jo Nye, "Physics and Chemistry: Commensurate or Incommensurate Sciences?" 205–44 in *The Invention of Physical Science*, eds. Mary J. Nye et al. (Dordrecht: Kluwer, 1992), on 217. Nye quotes from R. S. Mulliken, "Spectroscopy, Quantum Chemistry, and Molecular Physics," *Physics Today* 21, 1968, 52–57, on 55.

[52] Boerhaave, *New Method*, 1:173; Herman Boerhaave, *Elementa chemiae*, 2 vols. (Leiden, 1732), 1:79.

[53] Knoeff, *Herman Boerhaave*, 116–17. John Powers, *Herman Boerhaave and the Pedagogical Reform*, likewise argues that Boerhaave's pedagogical method was directed towards the development and analysis of theoretical claims, rather than the propagation of chemical remedies.

[54] CSH, 74.

[55] Herman Boerhaave, *Oration in qua repurgatae medicinae facilis asseritur simplicitas* (Leiden, 1709), 25; translated by Luyendijk-Elshout and Kegel-Brinkgreve in *Boerhaave's Orations*, 139–41.

[56] Hippocrates, *Aphorisms*, I.6; Boerhaave, *Kortbondige Spreuken*, 4.

[57] Gerard van Swieten, one of Boerhaave's influential pupils and commentators, first argued that Boerhaave "united the theories of the chemists and the mechanicians"; *Commentaries upon Boerhaave's Aphorisms Concerning the Knowledge and Cure of Diseases* (Edinburgh, 1776), x. Gerrit A. Lindeboom, "Boerhaave's Impact on the Relation between Chemistry and Medicine," *Clio Medica* 7, 1972, 271–78; "Boerhaave's Concept of the Basic Structure of the Body," *Clio Medica* 5, 1970, 203–08 also considered Boerhaave a mechanist. Luyendijk-Elshout, "Mechanicisme contra Vitalisme: De School van Herman Boerhaave en de Beginselen van het Leven," *Tijdschrift voor de Geschiedenis der Geneeskunde, Natuurwetenschappen en Techniek* 5, 1982, 16–26 does not take much notice of Boerhaave's chemistry for medicine. F.R. Jevons, "Boerhaave's Biochemistry," *Medical History* 6, 1962, 343–62, acknowledges (346) the importance of chemistry by stating that chemistry offered itself as the handmaid of medicine as a whole, not of pharmacy only. Allen Debus, *Chemistry and Medical Debate: Van Helmont to Boerhaave* (Canton: Science History Publications, 2001), however, has recently argued that although chemistry was relevant for Boerhaave's physiology, it was 'clearly not to become the theoretical basis of medicine' (201).

[58] CSEE, 211.

[59] Ibid.

BERNADETTE BENSAUDE-VINCENT AND CHRISTINE LEHMAN

PUBLIC LECTURES OF CHEMISTRY IN MID-EIGHTEENTH-CENTURY FRANCE

Through the eighteenth century chemistry became an autonomous science in many parts of Europe. In France, its promotion was indeed the result of the creation of a class for chemistry at the Paris Académie Royale des Sciences as early as 1699, whereas a class specifically for physics opened only much later in 1785. But the academic culture was only one facet of chemistry in the eighteenth century. Up to now most historians of chemistry have focused on the academic face of French chemistry. Consequently the emphasis has been put either on the famous episode of the Chemical Revolution conducted on the prestigious academic stage or, more recently, on the early eighteenth century when the Académie Royale des Sciences set up a program for the systematic investigation of plants.[1] The period in-between remains a black hole. The mid-eighteenth century looks like a forgotten period for chemistry, falling in the shadow of the star production of the French Enlightenment – Diderot's *Encyclopédie*. However, this alleged empty period was full of local events, such as chemistry courses. In an age when training and enlightening the public were not two separate activities, experimental demonstrations constructed chemistry as a fashionable and legitimate science. Not only did they craft many practitioners of chemistry but they were also compelling for philosophers such as Jean-Jacques Rousseau and Denis Diderot.[2]

This paper, focused on chemistry lectures and demonstrations, is an attempt to better understand how chemical knowledge was spread and transmitted. Chemistry is one of the few scientific disciplines whose educational component has attracted historical attention.[3] Hélène Metzger's classic survey of early modern chemical doctrines was based on the long series of textbooks published in the seventeenth century.[4] Then Owen Hannaway argued that textbooks were crucial for promoting chemistry as a science divorced from the hermetic tradition.[5] More recently a collective volume dedicated to the communication of chemistry presented a number of textbook traditions.[6] Yet, most of the time, the pedagogical component is perceived exclusively through textbooks. Apart from Rhoda Rappaport's classic articles on Guillaume-François Rouelle, oral didactic practices have been largely ignored, for the obvious reasons of a lack of accessible sources. However, recent studies based on archival materials suggest that Rouelle's demonstrations were but one example among a host of chemistry courses.[7] From a tentative survey of chemistry courses delivered both in Paris and in the provinces between 1750 and 1780, we try to identify the places where chemistry was taught, who the lecturers were, and, when

L. M. Principe (ed.), New Narratives in Eighteenth-Century Chemistry: Contributions from the First Francis Bacon Workshop, 21–23 April 2005, 77–96.

possible, who sat in their audiences. Then looking more closely at the contents of some of these lectures, we try to define the balance between practical applications and theoretical knowledge as well as the balance between experimental demonstration and discourse.

PLACES FOR CHEMISTRY

The panorama presented here as Table 1 does not claim to be exhaustive. Rather it is intended as a sample able to show the interplay of general trends and local circumstances that prompted the creation of chemical demonstrations. It is noticeable first

Table 1. Panorama of Some Chemistry Courses Delivered in France (1750–1780)[49]

(a) Paris

Town	Beginning in	Type of course	Location	Dates	Professor	Demonstrator or "assistant"[50]
Paris	1638	Public	Laboratory of the *Jardin des apothicaires*, Rue de l'Arbalette	1753–1768	In 1761: Couzié, Moret, Juliot, La Planche, Bataille, Santerre, Laborie, Azéma, Trevez[51]	
	1648	Public	Auditorium of the *Jardin du Roy*	1743–1768	Louis-Claude Bourdelin	Guillaume-François Rouelle
				1768–1771		Hilaire-Marin Rouelle
				1771–1779	Pierre-Joseph Macquer	
				1779–1784		
				1784–1793	Antoine-François de Fourcroy	Antoine-Louis Brongniart
	1771	Public	Medicine Faculty	1770–1776	Auguste Roux	Bellot + de La Planche
				1776–?	Jean-Baptiste Bucquet	Poissonnier (collège royal)
	1757	Private	Baumé's *officine*, Rue St Denis	1757–1773	Pierre-Joseph Macquer	Antoine Baumé
	1738	Private	Rouelle family's *officine*	1738–1768?	Guillaume-François Rouelle	
		Private		1768–1776	Hilaire-Marin Rouelle	
		Private	Demachy's *officine*	~1762–1768	Jacques-François Demachy	
		Private	De La Planche's *officine*	1751–1762	Laurent-Charles de La Planche	
		Private	De La Planche's *officine*	1780–?	Jean-Baptiste Bucquet	De La Planche
		Private	Brongniart's *officine*	~1778	Antoine-Louis Brongniart	
		Private	Mitouard's *officine*	~1776–1778	Mitouard	
		Private course of docimasy	Sage's *officine*	1760–~1776	Baltazar-Georges Sage	
	1778	Public	School of Mines	1778–?	Sage	

(continued)

Table 1 (continued)

(b) Provinces

Town	Beginning in	Type of course	Location	Dates	Professor	Demonstrator
Abbeville	1773 (?)	Private	De Ribaucourt's *officine*	1773–1779	Pierre De Ribaucourt	
Amiens	13 February 1777		Jacobins' Hall	1777–1778	Dhervillez	Alexandre-Ferdinand Lapostolle
	30 May 1778		Botanical Garden	1778–?	Dhervillez	Lapostolle
	1778	Private	Lapostolle's *officine*	1778–?	Lapostolle	
	1785	Private	Museum	1785–?	Lapostolle	
Angers	1786	Private	Tessié's laboratory	1786–1789	Tessié du Closeau	
Lille	1750	Private	Decroix's *officine*	1750?–1760	Louis-Joseph Decroix	
Metz		Private, with the support of the Société d'étude des Sciences et des Arts		1757–1758	Mathieu Peyevieux, Apothicaire major de l'hôpital militaire	Charles Claude Gervaise
		Private, with the agreement of the Société Royale des Sciences et des Arts de Metz	Thyrion's *officine*	1765–1769	Jean-Baptiste Thyrion	
	26 November 1778	Public, with the support of the Société Royale de Metz	Auditorium of the Société Royale de Metz in the Town Hall	1778–1790	Henry Michel du Tennetar	Jean-Baptiste Becoeur fils
Montpellier	1675–1676	Public	Auditorium of the Medicine Faculty	1732–1758	Antoine Fizes	Claude Thoynon
				1758–1765		Jacques Willermoz
				1765–1766	Gabriel-François Venel	
				1767–1771	Gaspard René	
				1771–?		Jean Joyeuse
	1761	Private	Montet's *officine*	1761–1767 or 70	Gabriel-François Venel	Jacques Montet

(continued)

Table 1 (continued)

Town	Beginning in	Type of course	Location	Dates	Professor	Demonstrator
Nancy	17 april 1776	Private	Nicolas's *officine*	1776	Henry Michel du Tennetar	Pierre-François Nicolas
		Private	Willemet's *officine*	1783–1784	Pierre-François Nicolas (physician)	Pierre-Remy Willemet
	15 april 1777	Public and charged	Nicolas's *officine*	1777–1778	Henry Michel du Tennetar	Pierre-François Nicolas
	1779		Medicine Faculty	1779	Henry Michel du Tennetar	Pierre-François Nicolas
		Public		1780–1782	Pierre-François Nicolas	Delaporte
				1782–?		Willemet
Rouen				1779–1780	Antoine-François Hardy	François-Antoine-Henri Decroizilles
	3 March 1777			1777–1781	Pierre-François Mesaize	
Strasbourg		Private	Spielmann's *officine* Pharmacie du Cerf in Strasbourg	1749–1759	Jacques Reinbold Spielmann	
	1685 Johann Boecler professor of chemistry and materia medica	Public	Medicine Faculty	1759–1783	Jacques Reinbold Spielmann	?
Toulouse	1705 Dufaur chaire de chimie et pharmacie	Public course of materia medica (chemistry and botanics)	Medicine Faculty	1775–1776	Louis-Guilllaume Dubernard	?

of all that chemistry was not a metropolitan affair. Although Paris certainly provided the largest audiences for chemical lecturers, many courses started in provincial towns, but generally later. Indeed not all of them were fully successful, yet the above chart cites a large number of enterprises that lasted for several years.

Another striking feature is that, in addition to the private fee-paying courses initiated in the seventeenth century that continued throughout the eighteenth century, a number of public courses were offered for free. It is clear that the urgency of chemistry for pharmaceutical and medical training was the foundation of most chemistry courses. The enduring alliance between chemistry and pharmacy, centered on the production of both medicines and other laboratory products prevailed all over the French kingdom. In Paris, for instance, chemistry courses were offered at the *Jardin des apothicaires* (the apothecaries' hall), a philanthropic teaching institution founded in the sixteenth century by Nicolas Houël (Figure 1). The Guild of Merchant Apothecaries and Spicers (*la Compagnie des Marchands Apothicaires-épiciers*) opened chemical lectures as part of the training for future apothecaries in 1700.[8]

The courses were held in the laboratory of the *Jardin des apothicaires* itself, which had been constructed in 1700 together with an amphitheater. The auditorium was an integral part in the laboratory, exclusively used for the teaching of chemical operations and preparations. An "Inventory of the House and Garden known by the name of the *Collège de Pharmacie*" drawn up in 1788 offers a very detailed description of the building:

> On the left we find a laboratory in which has been constructed a large oak tiered seating arrangement composed of nine tiers with two benches on the floor. The tiers are surrounded by a barrier.
>
> On the right of the fireplace there is a counter on which one can place the objects for the demonstrations, as well as the ovens and tools for the same ends ... At the back of the laboratory there is a mantle over the fireplace constructed along the whole of its length ... Under the aforementioned fireplace there are limestone supports for the ovens ... In the aforementioned laboratory there are fifty of the most ordinary chairs.[9]

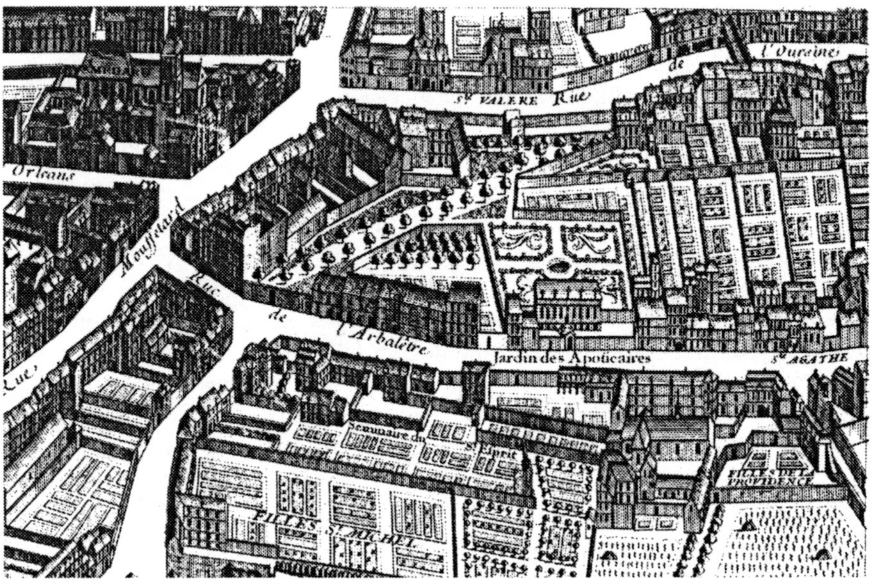

Figure 1. "Le Jardin des apothicaires" from a map by Louis Bretez commonly known as the "plan Turgot," 1739

The domain of the demonstrator – between the hearth of the fireplace and the table for the demonstrations – was distinct from that of the public, separated and protected by a barrier. The tiered seating allowed everyone to see and to follow the demonstrations. The mention of "fifty of the most ordinary chairs" suggests that the spectators' chairs were integral part of the laboratory equipment, along with the fireplace, the ovens, the glassware and the porcelain. In this place annual courses were delivered from 1700 to 1723. After an interruption, the courses re-opened in 1753, before being ended in 1768 by order of the Faculty of Medicine, unhappy with the idea of apothecaries pretending to the status of professors. The tradition of the *Jardin des apothicaires* is unique, since similar initiatives held by local corporations of apothecaries failed in Nantes in 1703, in Bordeaux in 1750 and Metz in 1753.[10]

The courses offered at the Paris *Jardin du Roy* (the royal botanical gardens), also derived from a pharmaceutical tradition. The *Jardin du Roy*, an official institution essentially dedicated to the collection of medicinal plants and founded in 1635[11] initially limited its teaching to botany. Nevertheless, chemistry succeeded in grafting itself onto this original teaching mission and came to represent an ever more significant proportion. Two positions, one as professor and the other as demonstrator were created to teach the composition of medicinal plants. The description of the professor's responsibilities changed several times, but always in the direction of an increasing preponderance of chemistry.[12]

In the provinces, botanical gardens were also a cradle for public chemical demonstrations. For instance, in the 1730s, the Strasbourg botanical garden dedicated a building to botanical and chemical demonstrations. The medical faculties in Strasbourg and Montpellier introduced chemistry courses in their curriculum, Montpellier in 1675, Strasbourg in 1685. But not all medical faculties had laboratory facilities for chemical demonstrations. For instance, the medical faculty created at Pont à Mousson in 1598 delivered pharmacy courses without experimental demonstrations for lack of laboratory facilities, until its transfer to Nancy in 1768.[13] In Toulouse, where a chair of chemistry and pharmacy had been created in 1705, a laboratory was not built until 1774, and a course of *materia medica*, chemistry, and botanics was delivered there in 1775 by Louis Guillaume Dubernard.[14]

In addition to the few medical faculties, there were a number of *Collèges de médecine* created by the medical Guild – in Lille, Lyons, Nancy, and Rouen – all places where chemistry courses were offered.[15] However, the historical connection among chemistry, pharmacy, and medicine should not hide chemistry's integration in the broader context of provincial cultural life. The local academies, whose number and influence on the French Enlightenment have been emphasized by Daniel Roche, prompted the culture of chemistry.[16] For instance, in Metz the *Société d'étude des sciences et des arts* founded in 1757 supported a private course of chemistry and hosted the public demonstration in its own auditorium when it became the *Société royale des sciences et des arts* in 1760. Even when the chemical demonstrations were not initiated by the local academy, a large proportion of the chemistry courses aforementioned took place in towns with local academies: Lyons had an academy founded in 1700, Arras in 1737, Amiens, the capital of Picardie north of Paris, had a science academy founded in 1745, Rouen in 1742, and Nancy had an *Académie des sciences et belles lettres* created in 1750, Besançon in 1752.[17] Learned societies, such as the *Société des Philathènes* created in 1759 in Metz by a group of gentlemen with literary and scientific interests, or the *Collège des philalèthes* created in 1785 in Lille, also offered propitious grounds for the implementation of chemistry courses. Ultimately, and unsurprisingly, the public for chemical demonstrations was predominantly found in provincial towns with a full educational system. Despite the existence of very few universities, many towns had Jesuit colleges, or *collèges de plein exercice*[18] or independent colleges as in Amiens, Abbeville, Metz, and Rouen. Indeed, there were no specific courses of chemistry in colleges, but teachers could include chemical topics and chemical demonstrations in courses of experimental physics. In short, in the mid-eighteenth century, urban centers with a proportion of educated people provided audiences for chemistry demonstrations that far exceeded the circles of pharmacists and medical doctors.

INCREASING AUDIENCES

The public for chemistry gradually increased and diversified. At the *Jardin des apothicaires*, free demonstrations were initially set up for professional training, and the public for these courses was above all composed of apprentice apothecaries who were joined by physicians, since the Faculty of Medicine of Paris did not offer its own chemistry courses. But the courses were widely publicized by means of 500 to 1000 posters pasted up in the streets of Paris. The archives mention 'a multitude of amateurs and of students from all states, both national and foreign who have come here to learn.' The increasing audience might explain why the laboratory needed to be enlarged in 1760.

As is already well known, Rouelle's demonstrations at the *Jardin du Roy* enjoyed considerable success. Until the renovation undertaken by Buffon in 1787, they were held in the auditorium in spring and summer where the anatomy course was delivered in the winter. With the rising temperature, the cadavers became difficult to preserve and so anatomy gave way to the botany and chemistry courses for the duration of the summer months, as advertised in the *Gazette de médecine* of 1761. Jussieu described the polyvalent auditorium in the following terms: "This amphitheater, which could hold 600 students, was located in the building that lay between the large entrance to the Jardin and the terrace of the great hillock." André Thouin, Buffon's gardener, added the information that "it was too small by half to contain the members of the audience."[19] Here then, a 600-seat auditorium was only half the required size! This suggests at least a 1000 people attending these courses – quite an impressive audience, even for a big city like Paris. According to Diderot, Rouelle's demonstrations attracted "a quarter of the city" from every class of society, including "the children of nobles who wanted to learn." Rouelle not only trained most of the chemists in the eighteenth century – Macquer, Venel, Brongniart, Bucquet, Sage, and Lavoisier, to mention only the best known of them – but also the *philosophes* of the Enlightenment: Diderot, Rousseau, Turgot, Malesherbes, as well as various other members of polite society.

Although they were not presented as shows for entertainment, chemical demonstrations attracted various social categories.[20] In fact, teaching and commerce were overlapping activities. Since the seventeenth century, many apothecaries had been delivering private courses in their shops, with experimental demonstrations of their art and commercialization of their products.[21] The *officine* was at the same time a place for the preparation of medicines and a place for commerce, being an annex of the pharmacy shop itself and a place for teaching. In the mid-eighteenth century, as commercial tourism and consumer society developed, those hybrid activities became increasingly fashionable.[22]

Remarkably, there was sufficient audience in the same town to have private courses alongside a public course, sometimes taught by the same demonstrator. Among the best known private courses were those taught by the two Geoffroy brothers, both master apothecaries and members of the Académie Royale des Sciences, with Etienne-François, the older brother, famous for his affinity table. They taught the course in their pharmacy shop in the rue Bourtibourg, while Rouelle taught in the rue Jacob from 1746, Macquer and Baumé in the rue St. Denis starting in 1757, and de La Planche in rue de la Monnaie. The "particular courses" were announced in medical journals, which suggests that they were primarily intended for medical audiences. Their number in Paris and the provinces is remarkable, since we know that the registration fees were very expensive. It cost 96 livres to follow Macquer and Baumé's course (half price for those who had already taken a private chemistry course). The course offered by Venel and Montet in Montpellier cost half that amount. How many people could afford such courses? We have no precise attendance records for such private courses, except one, and that might be inflated: in 1764, Venel claimed to have 42 people in his private course. Such figure suggests that the attendance included more than medical or pharmaceutical students. It is also remarkable that the number of private courses did not decrease in the 1770s when the medical faculties and the colleges of pharmacy took charge of the chemical education of medical doctors and pharmacists, respectively. One might think that this would have signaled the end for the private fee-paying courses, but this was not at all the case. Advertisements no longer appeared exclusively in the journals dedicated to medicine and pharmacy, as was the case 20 years before, but were now to be found in the daily papers. Although it is difficult to identify the audience of such courses, we should keep in mind that chemical experiments were of interest for various purposes. In fact, they were practiced widely throughout French society, not only by artisans, miners, glassmakers, dyers, and metallurgists for various professional productions, but also at home for domestic purposes. Despite the growing importance of the guild of perfumers in Paris, the women in many bourgeois families were still in charge of the fabrication of medicines, cosmetics, and cleaning products.[23] Their role required a familiarity with chemistry, as empirical practices had to be grounded on a minimum theoretical basis. Marie Murdrach's famous treatise *La chimie charitable et facile en faveur des dames* (1666) remained popular for decades.[24] In his colorful portrayal of everyday Parisian life, *Le tableau de Paris*, Louis Sébastien Mercier mentioned a variety of daily uses of chemistry, a science that he highly praised.[25] Moreover a number of aristocrats had their own private laboratories at home where they practiced a bit of chemistry for entertainment. For instance, the Fermier general (tax collector) Charles Dupin owned a private laboratory near Blois, and hired Jean Jacques Rousseau to teach chemistry to his son. Rousseau himself had been initiated into chemistry by Madame de Warens, who was fond of medical preparations.[26] He even authored a textbook of chemistry entitled *Institutions chymiques*, compiling various sources (such as Boerhaave, Rouelle, and Senac) to which Rousseau added his own personal reflections on chemistry.

PROFESSORS AND DEMONSTRATORS

Who were the people who delivered chemistry courses? While in the 1760s chemistry attracted wider audiences than just medical students, the people who delivered these chemistry courses, whether public or private, were mainly recruited from among pharmacists and medical doctors. Our survey confirms the strong dependence of chemistry on medicine and pharmacy in France, and at the same time it points to a number of new and revealing details.

At the *Jardin des apothicaires*, no single person held the teaching position. The courses were taught in turn by each master apothecary from the guild, each of whom was limited to teaching a maximum of two years in succession. The announcement for these courses presented the names of nine demonstrators charged alternately with the course in chemical experiments. De La Planche mentions that on 26 April 1769, the council of apothecaries appointed the following demonstrators: Guenaud, Cousière, Demorette, Picard, Julliot, La Planche, Bataille, Santerre, and Azema.[27] The contents were decided collectively, without any possibility for one of the lecturers to impose his own will.

At the *Jardin du Roy*, nearly all the chair holders as well as their replacements were members of the Paris Academy of Sciences, and all except Boulduc were doctors, while all the demonstrators (except for Davisson) were pharmacists. The situation of Guillaume-François Rouelle as demonstrator is unique in the history of the *Jardin du Roy*, as he was appointed with the title "demonstrator of chemistry at the *Jardin des plantes* under the title of professor of chemistry." The last demonstrator on the list, Antoine-Louis Brongniart had already been appointed demonstrator at the new *Collège de pharmacie* founded in 1777, when he joined the *Jardin du Roy* two years later.

As for private courses, the number of which far exceeded the public courses, most of them were founded by pharmacists in their *officines*. Pharmacists remained the main vehicles for chemical knowledge in France, which does not necessarily mean that the courses dealt exclusively with pharmaceutical preparations. For instance, for more than 15 years (1760–1776), Baltazar Sage delivered a course of *docimasie* (the quantitative analysis of metallic ores) in his *officine* until he was appointed to the National School of Mines.

The pharmacists who created courses in the provinces were one generation younger than the pioneers in Paris. In fact, most of them had acquired their knowledge and expertise in Paris. Lapostolle (1749–1831) who taught in Amiens did his "compagnonage" with Cadet de Vaux, a chemist and member of the Paris Academy of Science. Pierre Ribaucourt (1739–1806) who taught in Rouen had attended the courses of Rouelle, Macquer, and Baumé in Paris. Pierre-François Mesaize (1748–1811) had attended those of de La Planche's. Pierre François Nicolas (1743–1816) had served as soldier with Pierre Bayen and attended Rouelle's demonstrations in Paris. These profiles suggest that already in the eighteenth century, Paris was the center of scientific life and a *passage obligé* for a career in science.

Despite their crucial role in chemistry teaching, pharmacists did not enjoy a real leadership. In fact, the above table of courses shows that many courses were delivered by a duo made up of a pharmacist and a physician. Thus in Paris, the physician Pierre-Joseph Macquer taught with the apothecary Antoine Baumé; in Montpellier, Gabriel-François Venel, who was a physician, paired up with Jacques Montet; in the 1780s, the apothecary de La Planche started a new course with Jean-Baptiste Bucquet, a medical doctor. Courses could be taught by apothecaries working solo (such as Pierre de Ribaucourt in Abbeville or Jean Baptiste Thyrion in Metz) but medical doctors who wanted to deliver chemical demonstrations had to secure the partnership of an apothecary. The reason for such partnerships is provided by Félix Vicq d'Azyr in his *Eloge de Macquer*:

> Custom dictates that theory should be kept apart from demonstration and that these two aspects, which are mixed together in order to render teaching attractive, should be dealt with by two men, one of whom only talks, while the other acts and talks simultaneously.[28]

A footnote indicates that the custom of a professor and a demonstrator jointly teaching courses was also current in several universities in Germany and Italy. The origin of this custom in France lies in the statutes that governed the guilds. Medical doctors were required to teach in full costume and their lessons could not be other than '*scriptis et auribus*,' written or oral. They explicitly prohibited themselves from carrying out any manual operation, as they themselves were not supposed to get their hands dirty in sooty furnaces. As far as apothecaries were concerned, the statutes of their guild forbade anyone who was not a qualified apothecary from holding a demonstration. This prohibition applied particularly to medical doctors.

SUBVERTING SOCIAL AND EPISTEMOLOGICAL HIERARCHIES

This division of labor had a tremendous impact on the public perception of chemistry. It appeared as a knowledge torn rather than shared by two categories of experts. When delivered by two separate voices, theory and experiment might appear as disconnected rather than interdependent. The physical divorce between the carrier of theory and the performer of experiments encouraged the image of experiments as a kind of a pointless and amusing game. The *Gazette de médecine* concluded the announcement of a list of chemistry courses by the following deprecating remark:

> Our paper would not suffice to make known in detail all those who burn charcoal in Paris in order to illuminate such-and-such a chemical truth, and a full folio would hardly suffice just to name those who burn charcoal without knowing why they do it.[29]

At first glance, the pharmacist–physician duos tended to reinforce scholars' prejudices against laboratory work, since apothecaries occupied a socially subaltern position where physicians' organizations policed the preparation of drugs and inspected the pharmacies. Thus it comes as no surprise that the *Gazette de médecine* despised experimental chemists as people teasing fire for dubious purposes.

Nevertheless, chemistry courses undermined the well-established hierarchy of knowledge. First, doctors depended on the apothecaries if they wanted to open a private course, since they could not conduct the experiments themselves, whereas apothecaries enjoyed more independence since they could write a textbook accompanying their demonstrations. In fact, many courses went with treatises, and sometimes buying such a book was one of the preconditions for attending the course. Apart from Jean-Baptiste Bucquet, a medical doctor who published his course in 1771,[30] and one textbook jointly authored by the two partners in Amiens,[31] most textbooks published in the two decades 1750s to 1770s were authored by apothecaries: Antoine Baumé and Jacques-François Demachy, in Paris,[32] Louis-Joseph Decroix in Lille, Jacques Reinbold Spielmann in Strasbourg, and Pierre-François Nicolas in Nancy.[33] Second, whereas in public institutions professors received a higher salary than demonstrators, the contracts of partnerships for private courses specified that the pharmacist would receive more fees for performing the experiments and for supplying and maintaining the apparatus and chemicals.[34] So for a few years the private courses subverted the social hierarchy between physicians and pharmacists that was promptly re-established with the creation of chemistry chairs in the faculties of medicine. Third, the division of labor between speech and gesture, between discourse and demonstration, mind and hand, so to speak, could not be strictly applied. Apothecaries could hardly refrain from commenting on the experiments going on and thus developing their own views. They tended to operate across both fields, combining word and gesture, and eventually came to dominate the stage, relegating the physician to the role of 'narrator' behind the stage.

Gradually, the deprecating image of the manual worker, sweating in painful and slavish labors, was rivaled by an alternate rhetoric that praised experiment, as instantiated in Vicq d'Azyr's eulogy of Macquer. Although he was himself a medical doctor and a member of the Société Royale de Médecine, Vicq d'Azyr pointed out how his fellows were bogged down by this rigmarole.

> For a number of centuries physics has been nothing but a tissue of systems, a patchwork of authorities drawn from the ancients, that the doctors, fenced around with magisterial pomp, teach to their disciples. When the progress of knowledge forced them out of their schools to interrogate nature in the laboratory, they thought that to retain their dignity they needed to appear in their

> robes: these outfits mean they are reduced to the situation where it is impossible to do anything other than speechifying.[35]

Rhetorical defenses of experimental practices were flourishing in the 1750s, especially in the circles of *philosophes*. Witness Diderot's blazing offensive in *De l'interprétation de la nature* against the speculative and abstract knowledge of those who were able only to "reflect," and who "have many ideas and no instruments." In the same year 1753, Gabriel François Venel's heroic portrayal of the chemist as an "artist" in the article *Chymie* of the *Encyclopédie*, echoed Diderot's defense of experiment. Venel praised only the true chemist who was ready to take off the gown and base his knowledge on practical work:

> It is the necessity of all this practical knowledge, the length of chemical experiments, the assiduous labor and the observation that they demand, the expenses they occasion, the dangers to which they expose us, the eagerness to work at this kind of occupation that one is always in danger of contracting, that has caused the most sensible chemists to say that chemistry is a madman's passion.[36]

Far from despising experimental chemists for their insanity and eccentricity, Venel admired their courage and considered them "as citizens who deserve all our thanks."

Staunch advocates of chemical experiments were found mainly among Rouelle's former students. It seems that Rouelle's teaching did a great deal to change the social and epistemological models of the *bon savant*. In Rhoda Rappaport's classic monograph, Rouelle was a significant figure who deserved historical attention for spreading Stahl's theory in France; she stressed his creativity in pointing out that he re-invented Stahl's doctrine rather than simply transmitting it. But her Rouelle is a picturesque character and a brilliant doctrinaire. Here we want to suggest a different view of his importance in an effort to recapture what made Rouelle so attractive and fascinating. It was certainly not Stahl's theory which attracted hundreds of people to the auditorium at the Jardin du Roy and dozens to his private course on the rue Jacob.

Because he held the double position of professor and demonstrator, Rouelle's teaching was not coordinated with the theoretical discourse of the professor, Louis-Claude Bourdelin. Consequently Rouelle had to present both the chemical operations and the chemical theory simultaneously. He normally had two assistants, his brother Hilaire-Marin, and his nephew. Their role was to prepare the experiments for the demonstration. Sometimes, however, Rouelle performed the experiments himself, while talking, with the potential for explosions. Strikingly, all witnesses reported anecdotes of rather comic accidents due to Rouelle's difficulty to manage 'doing' and 'talking' at the same time. For instance, Friedrich Melchior Grimm wrote in his *Correspondance littéraire*, that one day Rouelle started igniting an essential oil with spirit of niter, then left the experiment alone for a moment to finish his explanation while turning to face the audience.

> Suddenly, the ignition experiment exploded and broke the lid with a crack, giving off a bright light and filling the amphitheater with thick and suffocating smoke. The terrified public immediately started to flee and fanned out through

> the garden in fear, while the operator, stunned and motionless, escaped with only the loss of his wig and shirt-cuffs.[37]

In Diderot's recollections, one day Rouelle was demonstrating a distillation process. He remarked to his audience,

> "Gentlemen, do you see this cauldron on the fire? Well, if I stopped stirring it a single instant, an explosion would occur that would blow us all up!" While saying these words, he did not fail to forget to stir it and his prediction came to pass: the explosion occurred with a terrific noise, shattered all the glass windows of the laboratory, and, in no time, two hundred students found themselves scattered in the garden.[38]

As Lissa Roberts emphasizes, all testimonies suggest that Rouelle's public demonstrations were theatrical performances.[39] And this theatrical – spectacular – way of transmitting chemical knowledge may be the chief reason for Rouelle's immense popularity.

REVISITING OUR CATEGORIES

In order to assess the respective values of theory, experiment, and practice in mid-eighteenth-century French chemistry courses, it is wise to avoid using the conventional view of experiment as an illustration or proof of a theoretical point that later came to prevail in the pedagogy of experimental science. Similarly, it would be inappropriate to frame practical aspects in terms of applied theory since the division between pure and applied chemistry was far from stabilized in the mid-eighteenth century.[40]

How are we to understand the function of experiments in chemistry courses? A look at the announcements of Rouelle's private courses suggests that experiment was above all a show. It was under the title "experiments" that Rouelle advertised his courses. He promised spectacular effects with a minimum theoretical baggage.

> With these experiments we will limit ourselves to making known the advantages that physics and medicine have drawn from works of chemistry. Further, we will make every effort to give examples of the utility of these same operations in several arts, and even their utility in everyday domestic uses.[41]

Rouelle downplayed the role of theory while he promised a great, sensational show. The last part of his course would reveal "the substances that are taken from the entrails of the earth" and would be the object of "unusual experiments" on bitumen, niter, marine salts, and the acids. The effects of all the mixtures would produce "changes in colour, detonations and the production of flames!"

It was not just Rouelle who used "experiments" as advertising slogans. Most of the private teachers tended to highlight the same two features of chemistry: experimental shows and practical applications. Rouelle covered such topics such as the embalming of mummies, Venel devoted parts of his lectures to the production of fine wines and good perfume, or criticisms of other public figures. De La Planche, who taught his own private course as well as teaching at the *Jardin des apothicaires*, and the Faculty of Medicine offered a veritable technical training. He also promised "curious chemical experiments" on metals, as well as experiments on the theme of "the discoveries made by some of the most famous chemists in Europe." Unlike Rouelle, he subtly emphasized theory, even including it in his title "A course of experimental chemistry, following the principles of Becher, Stahl, and Boerhaave."[42] He started with theory, then, while studying the vegetable kingdom he presented technical operations such as maceration, infusion, decoction, and so forth. Wherever possible, he would put the emphasis on applications, such as the "art of the bulk treatment of ores."

It seems then that chemical demonstrators favored the image of a useful science, rather than the image of an amusing science. Significantly, in Diderot's *Encyclopédie* there is no entry for "récréations chimiques" although Venel had announced it in his famous article *Chymie*.

> The most curious and magical branch of natural magic is that which works its wonders by chemical agents and upon chemical subjects. Phosphoruses, the inflammation of oils with acids, fulminating powders, violent effervescences, artificial volcanoes, the production, destruction, and sudden changes of color in certain liquids, unexpected precipitations and coagulations, etc. – even while neglecting the evidently chimerical claims of the divine Stone, rejuvenations, the homunculus of Paracelsus, the miracles of palingenesis, etc. – all these marvels I say, even in this enlightened age, can still astonish most people, or at least amuse them. See *Récreations chimiques*.[43]

The spectacular effects generated by chemical experiments were perhaps too close to magic and Venel did not want to favor such an area in his effort to promote chemistry as an integral part of natural philosophy.

As mentioned above, the training of pharmacists and physicians was the prime mover behind the creation of public and private chemistry courses. This style of pedagogy contrasted with traditional teaching methods. In their apprenticeship, artisans acquired their manual skills, their 'habitus,' through actual practice. Physics or natural philosophy was usually taught in the last year of secondary colleges, along with logic, ethics, and metaphysics, in a magisterial way or by reading a treatise. In contrast, students attending chemistry courses were supposed to acquire notions about the therapeutic virtues of individual substances and skills in the art of pharmaceutical preparations, through seeing experimental performances and listening to related comments. The success of such demonstrations relied on the conviction that knowledge enters the mind through the senses. As pointed out by Jessica Riskin, sensualist philosophy underlay the sensationalism of public demonstrations in the eighteenth century.[44] In physics, however, the experiments performed by demonstrators such us Willem 'sGravesande or John Theophilus Desaguliers were perceived as retranslations of general laws into experiments, whereas in chemistry, experiments instead revealed a specific operation using both natural agents and artificial instruments.

The vocational aim first determined the kind of theory that would be developed in such courses. It was not exactly the right place for discussing matter theory or rival views of its elemental components and how they combine. Such topics were not excluded, but they were confined in one or two lessons or occasionally developed in passing. Most chemistry teachers distrusted systems and advocated a Baconian approach to nature, partly out of necessity to adjust supply to demand. According to Baumé, his courses included more than 2000 experiments, through which he would analyze the three kingdoms of nature. In view of the many notebooks of Rouelle's lectures, his presentations consisted mostly of experiments. He confined theoretical statements to a few introductory remarks, then performed and described many procedures: 56 for plant chemistry, 11 or 12 for animal chemistry, and 159 or 160 for mineral chemistry. The disproportion between theory and experiment may also indicate that experimental procedures were a top priority for the students. Venel's students carefully noted the tricks and skills, the *tours de mains* that were probably mentioned by the demonstrator during the performance. However, many of them also noted the theoretical interpretations of the procedures described. Seldom or never were experiments presented as "translations" of the theoretical views expressed by the speaker into factual and sensual evidences. Experiments were an end rather than a means.

Displaying procedures, transmitting skills and manual know-how, was only one function of experiments. They also served a purpose of validation: analysis and synthesis were performed to advance the knowledge of natural compounds. For example, the combined actions of distillation and solvents were used to access to the composition of plants and animal substances, such as blood or bile. The preparation of various salts (acids, alkalis, and salts), were used to show the properties of mineral substances such as metals and non-metals, as well as their applications. What could be the structure and function of theory when it was basically only a flow of words accompanying gestures and colorful displays that kept sight, smell, hearing, and even touch on alert, as is suggested by many details in students' notebooks?

The comments about what was happening on the stage were not meant to provide a coherent system with an account of the ultimate causes acting in nature. Neither Rouelle, nor Venel, nor even Macquer cared to teach and spread any doctrine about chemistry. This does not mean that experimental demonstrations were theory-free. Rather, a theoretical discourse was needed for making sense of what was going on the bench before the spectators. In the first lesson of his course, Venel explicitly assumed that there was a complementarity between theoretical views and manual practices. Chemistry, he said, is equally divided into "fundamental" chemistry and detailed chemistry (*chimie de détail*); neither can exist without the other:

> ... when joined in the same subject, they establish the perfection of the science. The latter without the former is nothing but a handicraft, an ornament for the memory, the ability to carry out particular chemical acts, often very useful ones it is true, which do very well to be guided by the science. Likewise, fundamental scientific chemistry is so much the less perfect as it is more deprived of the help of the latter. For the true chemist almost never carries out even the most common operations without drawing from his observation of the phenomena some new notions that extend or correct fundamental truths.[45]

The continuous flow of speech was like a narrative identifying the actors and various protagonists of the chemical performance far beyond the realm of sensual data. For instance, Rouelle used the notion of element as a key for interpreting what was happening in the process. Taking up Stahl's concept of phlogiston, Rouelle made a decisive conceptual shift. He rejected the ancient distinction between elements (ultimate molecules) and principles (first compounds made of elements). He nevertheless admitted four elements: air, earth, water, and fire. He thus simultaneously redefined Stahl's phlogiston as the Aristotelian element of fire and rejuvenated the four-element theory in the light of affinity chemistry and displacement reactions. Thus the importance of this ancient doctrine in the mid-eighteenth century was not a reverence to the past, a marked traditionalism among chemists, as Duncan has suggested.[46]

Neither Rouelle nor his fellow chemistry teachers minded taking their theoretical views from a variety of sources. They gladly borrowed basic notions from Stahl – specifically his organization of matter and the term "phlogiston" to designate elemental fire – and combined them with Newton's views on affinities as attractions, and mixed all that with views taken from Boerhaave's works. In this respect the facile identification of mid-eighteenth-century French chemistry as "Stahlian" appears as a post-Lavoisier reconstruction of the past. The chemistry taught in Paris, Montpellier, Lyons, or Lille was neither "Stahlian" nor "Newtonian."

Looking closely at the notebooks of Rouelle's students in Paris, or Venel's students in Montpellier, it is clear that the distinction drawn up by Metzger between three different trends of chemistry influenced by Stahl, Newton, and Boerhaave is an artificial and abstract organization of chemical knowledge resulting from a quest for influences. Similarly, Pierre Duhem's distinction between various "schools of thought" – Cartesian, Newtonian, and empirical – overlooked the baroque effervescence of French chemistry in the mid-eighteenth century.[47]

In particular, Rouelle adopted Boerhaave's view of elements as constituents and instruments and developed a sort of operationalist matter theory. He presented his four-element theory under the heading of "instruments." This lesson included four natural instruments – fire, air, water, and earth – and two artificial instruments – menstrua and vessels. Rouelle attributed a dual function to elements; they were both the constituent units of mixts responsible for the conservation and transport of individual properties through chemical changes, and they were instruments of chemical reactions. The ancient radical distinction between nature and human artifacts was thus blurred in favor of an instrumental view of matter as an active process of operations. Material principles were always at work, circulating from mixts to mixts whether it be in the chemist's vessels or in the depths of the earth and the heights of the heavens. Rouelle's elements were individual, indestructible, and radically invisible, never isolable. They abandoned one combination to enter into another mixt. Thus they were made accessible only through displacement reactions, through the chemists' operations performed in the laboratory. In stark contrast to corpuscularian theories of matter, principles were not characterized as ontological units. Rather they were defined as units of operations *of* nature and *on* nature. One single word – operations – was used for what we presently call reactions and manipulations; hence, there was a specific mode of theorizing quite different from theories of physics. The chemical theory was framed by operating priorities as a discourse meant to make sense of practical operations.

Venel's courses in Montpellier reinforced the priorities of both experiment and practice over theory in two respects. First, in stark contrast to the polemical views developed in his famous article *Chymie* in Diderot's *Encyclopédie*, where Venel fired criticisms at Descartes and Newton and sanctified Stahl, his course in Montpellier developed an experimental approach based on affinity tables. He still advocated Stahlian views but combined them with those of Boerhaave and Newton. Second, whereas Rouelle confined his theoretical views to two introductory lessons – entitled "des généralités de la chymie" and "des principes chimiques," respectively – Venel did not approach theoretical topics until the end of the course. He developed his views on the four elements, their nature and roles in chemical phenomena in the sixty-second lesson. Far from engaging in discussions about rival theories, he simply summarized the major points to be memorized, such as the element water should be distinguished from the notion of fluidity, the element air enters as a constituent into various bodies, elemental "earth" is just a conjecture, and fire can be viewed either as an element (when it is called phlogiston) or as instrument, and "artist" chemists should be extremely careful in handling the latter. These theoretical remarks were not, however, his last word. Venel added a 63rd lesson on menstrua, in which he recapitulated the didactic uses of affinity tables, and ended up finally with a sixty-fourth lesson entitled "tableau des opérations," that surveyed all the chemical operations encountered along the course. Far from portraying chemistry as a consistent system, a branch of natural philosophy competing with physics for understanding matter, as he did in the article *Chymie* of the *Encyclopédie,* Venel the teacher presented chemistry a theater of operations managed by artists.

CONCLUSION

To what extent does this quick survey of a number of chemistry courses offered to the French public in the mid-eighteenth century shed new light on chemistry in the Enlightenment? First, in addition to the academic practice of chemistry on the stage of the Paris Royal Academy of Sciences that has retained the attention of many historians of chemistry, there was another culture of chemistry as a public science. "Public" here means that chemistry was widely spread in the public sphere, present in many towns and a part of urban culture. Second, utility was the prime mover for the establishment of such courses. The majority of the public was interested in receiving professional training, and we need to remember that chemistry existed only as a service science at the time, an auxiliary to medicine and pharmacy. Yet chemistry attracted wider audiences. A heterogeneous public was characteristic of all the chemistry demonstrations around the middle of the eighteenth century, as chemistry became part of polite culture. Chemical demonstrations were presumably less entertaining than the electrifying shows delivered by the Abbé Nollet and dozens itinerant lecturers, yet public and private teaching enterprises operated side by side, and a number of chemical demonstrations were a combination of pedagogical and commercial enterprises. Third, French mid-eighteenth-century chemistry can be only partially identified with the heroic science that Venel portrayed in his article *Chymie* for Diderot's and D'Alembert's *Encyclopédie*. Indeed, it was a knowledge based on sensory experience, tangible and visible phenomena, but it was not a Stahlian and anti-Newtonian science.[48] More generally, the distinctions between Cartesian, Newtonian, and Stahlian schools that have generated scholarly disputes ignore the fact that chemistry lecturers were not affiliated to any specific doctrine. Fourth, experimental demonstration was an end in itself rather than the foundation or the illustration of theory. The function of theory was not so much to "save phenomena" as to organize them into a teachable subject. Finally, chemistry lecturers deliberately blurred the ancient boundary between nature and artifact which prevailed in medieval scholastic culture. They thus favored the emergence of a new value system that praised public utility and technological operations with nature and upon nature.

NOTES

1 For a survey of the abundant literature on the Chemical Revolution see Patrice Bret, "Débats et chantiers actuels autour de Lavoisier et de la révolution chimique," *Revue d'Histoire des Sciences* 48, 1995, 3–8. F. L. Holmes drew the attention of historians of chemistry to the early days of the Paris Académie Royale des Sciences in *Eighteenth-century Chemistry as an Investigative Enterprise* (Berkeley, CA: Office for the History of Science and Technology, 1989), "Analysis by Fire and Solvent Extractions: The Metamorphosis of a Tradition," *Isis* 62, 1971, 129–48, "The Communal Context for E. F. Geoffroy's 'Table des rapports,'" *Science in Context* 9, 1996, 289–311; Rémi Franckowiak and Luc Peterschmitt, "La chimie de Homberg: une vérité certaine dans une physique contestable," *Early Modern Science and Medicine* 10, 2005, 65–90.

2 Rousseau, *Institutions chimiques*, Diderot, *De l'interprétation de la nature*, 1753.

3 On the relative absence of pedagogy on the map of science see David Kaiser's introduction to the collective volume *Pedagogy and the Practice of Science, Historical and Contemporary Perspectives*, ed. David Kaiser (Cambridge, MA: MIT Press, 2005).

4 Hélène Metzger, *Les doctrines chimiques en France du début du XVIIe siècle* (Paris: Les Presses Universitaires, 1923).

5 Owen Hannaway, *The Chemists and the Word: The Didactic Origins of Chemistry* (Baltimore: Johns Hopkins University Press, 1975).

6 Anders Lundgren and Bernadette Bensaude-Vincent, eds. *Communicating Chemistry: Textbooks and their Audiences, 1789–1939* (Canton, MA: Science History Publications, 2000).

7 John Perkins, "Creating Chemistry in Provincial France before the Revolution: The examples of Nancy and Metz, Part 1: Nancy," *Ambix* 50, 2003, 145–81, "Part II: Metz," 51, 2004, 43–75. Thanks to Prof. Pierre Labrude for indicating this article. See also Pierre Labrude, "Les débuts de l'enseignement de la chimie en Lorraine ducale: le Collège royal de médecine et la Faculté de médecine de Nancy (1756 et 1776)," *Revue d'histoire de la pharmacie* 53, 2005, 199–220. Gilbert Dalmasso, *Présence de la chymie dans la France du Nord, de la deuxième moitié du XVIIIe siècle au premier tiers du XIXe siècle*, PhD Université Charles de Gaulle de Lille III, 2005; Christine Lehman, *Gabriel-François Venel (1723–1775): Sa place dans la chimie française du XVIIIe siècle*, PhD Université Paris X-Nanterre, 2006.

8 "[E]very year, one of the apothecaries from the aforementioned guild will offer a free, public chemistry course for the instruction of those who practice medicine and pharmacy;" archives of the Faculté de pharmacie de Paris, register 37, fol. 58r, quoted by Gustave Planchon, *L'enseignement des sciences physico-chimiques au Jardin des apothicaires et à l'Ecole de pharmacie de Paris* (Paris: Flammarion, 1897),

12. A fund was created "pour subvenir aux frais que l'on peu faire pour la demonstration du cours de chymie quelle (la Compagnie) desire qu'on fasse tous les ans," fol. 59r.

[9] Archives of the Faculté de pharmacie de Paris, register 43, 19–20; "Etat de la Maison et Jardin appelé Collège de Pharmacie" par M. Essart Me Maçon demeurant rue St Etienne près St Etienne du Mont; à la réquisition de Mrs Bataille et Solomé prévôts et du Sr Santotte Ecrivain déchiffreur.

[10] Perkins, "Creating Chemistry, Part II," 43–44; G. Courteix, *Contribution à l'histoire de la pharmacie à Nantes: Le Jardin des apothicaires* (Baugé: Imprimerie du pays beaugois, 1929).

[11] Prior to the official edict creating the garden, Guy de la Brosse, one of Louis XIII's physicians, received permission for creating a royal botanical garden in 1626, and the land in 1633.

[12] Jean-Paul Contant, *L'enseignement de la chimie au jardin royal des plantes de Paris* (Cahors: A. Coueslant, 1952).

[13] Eugène Martin, *L'Université de Pont à Mousson (1572–1768)* (Paris and Nancy: Berger-Levrault and Co., 1891).

[14] Jules Barbot, *Les chroniques de la Faculté de Médecine de Toulouse du XIII au XXe siècle*, 2 vols. (Toulouse: Adolphe Trinchant, 1905).

[15] In Lille, the military hospital founded in 1774 hosted chemistry courses.

[16] Daniel Roche, *Le siècle des Lumières en province: Académies et académiciens provinciaux (1680–1789)*, 2 vols. (Paris: Editions de l'EHESS, 1989), 373. Roche noticed an increasing number of chemical subjects in provincial academies in the decade 1760–69.

[17] Ibid., 2:191–93.

[18] The "collèges de plein exercice" were in charge of the last two years of secondary school. The two years of "philosophy" included logic, metaphysics, ethics, mathematics, and physics.

[19] Bibliothèque centrale du Muséum national d'histoire naturelle, MS 1934, third notice, on 6, note(s). The sixth notice, on 17, states "L'amphithéâtre ancien situé entre la cour et une rue très passagère étoit trop resseré pour le nombre des élèves qui venait assister aux divers cours et les leçons étoient interrompues par le bruit des voitures. Il étoit dans le bâtiment qui existe entre la porte d'entrée et la terrasse de la grande butte."

[20] Larry Stewart, *The Rise of Public Science: Rhetoric, Technology, and Natural Philosophy in Newtonian Britain, 1660–1750* (Cambridge, Cambridge University Press, 1992); Simon Schaffer, "Natural philosophy and public spectacle in the eighteenth century," *History of Science* 21, 1983, 1–43.

[21] Michel Bougard, "Cours et demonstrations de chimie en France au XVIIe siècle," *Scientarium Historia* 19, 1993, 29–41.

[22] C. Walsh, "Shopping et tourisme: l'attrait des boutiques parisiennes au XVIIIe siècle," 223–38, in N. Coquery ed., *La boutique et la ville: Commerces, commerçants, espaces et clientèles XVIe–XXe siècle* (Tours: CEHVI-Université François-Rabelais, 2000).

[23] See, for instance, Catherine Lanoe, "Les jeux de l'artificiel: Culture, production et consommation des cosmétiques à Paris sous l'Ancien Régime XVI–XVIIIe siècle," PhD, University of Paris I, 2003.

[24] Marie Murdrach, *La Chymie charitable et facile en faveur des dames* (Paris, 1666; reprint ed, Paris: CNRS éditions, 1999).

[25] For instance, Louis Sébastien Mercier, *Le tableau de Paris* (Paris: Éditions la découverte, 1998), 226 pointed out that young maids who had lost their virginity before marriage used vinegar manufactured by Monsieur Maille in order to save the appearances on their wedding night. He added the ironic comment that "tout est régénération devant les lois de la chimie; la félicité des époux est encore liée à cette science sublime que j'idolâtre; elle fait la gloire, le bonheur et le repos des demoiselles parisiennes."

[26] Rousseau, *Confessions* (Paris: Garnier Flammarion, 1968), 88; Bernadette Bensaude-Vincent and B. Bernardi, "Rousseau chimiste," in Bensaude-Vincent and Bernardi, eds. *Rousseau et les sciences* (Paris: Lharmattan, 2003), 59–76.

[27] Archives de la Faculté de pharmacie de Paris, register 38, fol. 20v.

[28] Felix Vicq d'Azyr, "Eloge de Macquer," *Histoire de la société royale de médecine, 1782–1783* (Paris: Théophile Barrois, 1787), 74.

[29] *Gazette de Médecine* 1761, 199–200.

[30] Jean-Baptiste Bucquet, Introduction à l'étude des corps naturels, tirés du règne minéral (Paris, 1771). Indeed, Pierre-Joseph Macquer, a medical doctor, also authored two famous textbooks, *Elemens de*

chymie théorique (Paris, 1749) and *Elemens de chymie pratique* (Paris, 1751). However, when he opened a private course with Antoine Baumé, they co-signed a *Plan d'un cours de chymie expérimentale et raisonnée avec un discours historique sur la chymie* (Paris, 1757).

[31] Dhervillez & Lapostolle, *Plan d'un Cours de chymie expérimentale raisonnée et appliquée aux Arts* (Amiens, 1777).

[32] Antoine Baumé, *Manuel de chymie, ou Exposé des opérations et des produits d'un cours de chymie* (Paris, 1763); Antoine Baumé, *Chimie expérimentale et raisonnée* (Paris, 1773); Jacques-François Demachy, *Instituts de Chymie ou principes élémentaires de cette science, présentés sous un nouveau jour* (Paris, 1766); Jacques-François Demachy, *Procédés chymiques, rangés méthodiquement et définis … pour servir de suite aux Instituts de chymie* (Paris, 1769).

[33] Louis-Joseph Decroix, *Physico-chymie théorique en dialogue* (Lille, 1768); Jacques Reinbold Spielmann, *Institutiones chemiae praelectionibus academicis adcommodata, Instituts de chymie, traduit par Cadet le jeune* (Paris, 1770); Pierre-François Nicolas, *Cours de chymie théorico-pratique à l'usage des etudians et amateurs* (Nancy, 1777); Pierre-François Nicolas, *Précis des leçons publiques de chimie et d'histoire naturelle* (Nancy, 1787).

[34] For instance, Perkins reports that in Metz, Michel du Tennetar was to receive one third of the fees while the pharmacist Nicolas was to receive two-thirds, "Creating Chemistry, Part II," 166.

[35] Vicq d'Azyr, "Eloge de Macquer," 74. Here he evokes the functioning of the courses at the Jardin du Roy, but this applies equally to private courses.

[36] Gabriel-François Venel, "Chymie," Denis Diderot and d'Alembert, eds., *Encyclopédie ou Dictionnaire raisonné des sciences, des arts et des métiers* (Paris, 1753), 3:421a.

[37] Paul-Antoine Cap, *Guillaume-François Rouelle: Biographie chimique* (Paris, 1842), 17 and 23–4.

[38] Ibid., see also Friedrich Melchior Grimm, *Correspondance littéraire, philosophique et critique par Grimm, Diderot, Raynal, Meister, etc.*, ed. Maurice Tourneux, 16 vols. (Paris: Garnier, 1877–82), 9: 108; " 'Vous voyez bien, messieurs, ce chaudron sur ce brasier? Eh bien, si je cessais de remuer un seul instant, il s'ensuivrait une explosion qui nous ferait tous sauter en l'air!' En disant ces paroles il ne manqua pas d'oublier de remuer et sa prédiction fut accomplie: l'explosion se fit avec un fracas épouvantable, cassa toutes les vitres du laboratoire, et, en un instant, deux cents auditeurs se trouvèrent éparpillés dans le jardin."

[39] Lissa Roberts, "Chemistry on stage: G.F. Rouelle and the theatricality of Eighteenth-Century Chemistry," B. Bensaude-Vincent and C. Blondel, eds., *Science and Spectacle* (London: Ashgate, forthcoming).

[40] Christoph Meinel, "Theory or Practice? The Eighteenth-Century Debate on the Scientific Status of Chemistry," *Ambix* 30, 1983, 121–32.

[41] Bibliothèque nationale de France, document numérisé S 6436: Annonce du cours privé de Rouelle "en sa maison" rue Jacob et débutant le 17 novembre 1766.

[42] *Gazette de Médecine* 1762, 342–43: Announcement of de La Planche's course.

[43] Venel, "Chymie," 420b.

[44] Jessica Riskin, *Science in the Age of Sensibility: The Sentimental Empiricists of the French Enlightenment* (Chicago: University of Chicago Press, 2002).

[45] Wellcome Institute Library, London, MS 4914 [Notebook by Balme]: "Cours de Chymie fait chez monsieur Montet apoticaire par monsieur Venel Docteur et professeur en L'université De medecine à Montpellier, 1761," 3.

[46] Alistair W. Duncan, *Laws and Order in Eighteenth-Century Chemistry* (Oxford: Clarendon Press, 1996).

[47] Pierre Duhem, *Le Mixte et la combinaison chimique* (Paris: Fayard, 1985); Hélène Metzger, *Newton, Stahl, Boerhaave et la doctrine chimique* (Paris: Albert Blanchard, 1930).

[48] See, for instance, Guédon Jean-Claude, "Chimie et matérialisme: la stratégie anti-newtonienne de Diderot," *Dix-huitième siècle* 11, 185–200.

[49] The courses of physics taught in secondary colleges are omitted in this survey although some of them included chemistry.

[50] Demonstrator is designated by royal authority; this title corresponds to a public course chair.

[51] Alternating professors.

URSULA KLEIN

APOTHECARY-CHEMISTS IN EIGHTEENTH-CENTURY GERMANY

INTRODUCTION

In 1784, a connoisseur of intellectual life in Berlin reported to Lorenz Crell, professor of theoretical medicine and *materia medica* at the University of Helmstedt and editor of the *Chemische Annalen*, that interest in chemistry had grown enormously in his city: "You will hardly believe how much the study of chemistry is appreciated here now," he exclaimed. "Lectures on chemistry are attended by people from all social classes (*Stände*); what's more, since this winter there are also distinguished members of the fairer sex in the audience." These ladies, he continued, "forsake their coffee and game tables, their assemblies and picnics, to staunchly endure cold and heat, fumes and charcoal dust, and all other discomforts of a chemical workshop."[1] Crell was ready to believe this observation. Since 1783, Martin Heinrich Klaproth's public lectures on chemistry had indeed attracted a large audience in Berlin, among them many ladies.[2] Furthermore, the impressive number of 424 subscribers to his own journal in the very same year was not least a manifestation of chemistry's success in Germany.[3] In 1778, when Crell set out to publish the first issue of his chemical periodical, he emphasized that "chemistry's extended influence on learning and its great utility for the commonweal are so generally recognized that they need no proof."[4] In Crell's view, Germany was other nations' "acknowledged teacher" of chemistry. "Nature itself," he proclaimed, "seems to have destined us to become chemists; and, as we fulfill this calling, there is perhaps no other country where there are so many chemists (*Scheidekünstler*), be they true or false, as in Germany."[5]

Chemistry was indeed thriving in eighteenth-century Germany. Between 1720 and 1780 the number of renowned German chemists had more than doubled.[6] About half of these well-known German chemists were introduced to chemistry as medical students, and among the remaining half, the vast majority became acquainted with chemistry as pharmaceutical apprentices and practicing apothecaries.[7] As metallurgy was another important field for learning chemistry, many chemists also visited mines, smelting works and foundries, taking courses in mineralogy and assaying.[8] In eighteenth-century Germany, like other European countries, apothecaries were artisans who, as a rule, were apprenticed in an apothecary's shop.[9] But they also often carried out chemical investigations. As Karl Hufbauer pointed out, among the 564 German subscribers to Crell's *Chemische Annalen* between 1784 and 1789, 260 (46%) were apothecaries, and

L. M. Principe (ed.), New Narratives in Eighteenth-Century Chemistry: Contributions from the First Francis Bacon Workshop, 21–23 April 2005, 97–137.

among its contributors more than 40% were apothecaries as well.[10] The apprenticed and practicing apothecaries were even the most active contributors to Crell's journal, with single contributors publishing up to 68 papers during the five-year period from 1784 to 1789.[11] Among the nine leading German chemists around 1780 – L. Crell, J. F. Gmelin, F. A. C. Gren, S. F. Hermbstädt, M. H. Klaproth, J. F. Westrumb, J. C. Wiegleb, F. C. Achard, and J. F. A. Göttling – six (Gren, Hermbstädt, Klaproth, Westrumb, Wiegleb, and Göttling) were apprenticed apothecaries, and Wiegleb and Westrumb remained practicing apothecaries throughout their professional career.

The following study is concerned with the hybrid persona of the apothecary-chemist in eighteenth-century Germany. How did apothecaries learn chemistry? How did they become renowned chemists? What are the characteristics of the persona of an apothecary-chemist in eighteenth-century Germany? Questions like these are central to our historical understanding of eighteenth-century chemistry and its disciplinary boundaries, with respect not only to Germany but also to other countries, including France.[12] Eighteenth-century chemistry was far from being unambiguously a science, teaching discipline, and academic culture, but also reached out to technology and production, both before and after the Chemical Revolution.[13] Accordingly, many eighteenth-century chemists were hybrid savant-artisans that defy our distinction between eighteenth-century scholars, Enlightenment savants, and philosophical chemists on the one hand, and artisans and technological chemists on the other. Our narrow focus on the theoretical upheavals in the period of the Chemical Revolution in France has all too often blinded us to the interconnectedness of chemical science, technology, and production throughout the eighteenth century and well beyond. Early modern learned polemics against manual labor, and the rhetoric of Lavoisier and his associates, have further distorted our picture of the relationship between early modern practitioners and learned men (later "scientists") in general, and between apothecaries and chemical philosophers in particular. The contribution of apothecaries, metallurgical officials, mining councilors, chemical manufacturers, chemical–technological inventors and other skilled practitioners to the development of chemical science has not yet been studied in sufficient historical detail.

Instead of presenting a comprehensive account of chemistry in eighteenth-century Germany, my study illuminates some of its neglected aspects: the interconnectedness of the pharmaceutical handicraft and chemistry, and the hybrid persona of the apothecary-chemist.[14] In the first part of my paper I will describe the system of pharmaceutical apprenticeship in eighteenth-century Germany, with a specific focus on the question of how much chemistry was involved in it. How did pharmaceutical apprentices and journeymen learn chemistry in eighteenth-century Germany? My overview will show that the tradition of pharmaceutical apprenticeship in a local apothecary's shop was in a state of dissolution, in accordance with the innovations engendered in the entire pharmaceutical art through the proliferation of chemical remedies during the seventeenth century. In the second and third parts I will examine additional social institutions for apothecaries' learning of chemistry and for their active contribution to chemical science. Traveling, during the time of service as journeyman or later as owner of an apothecary's shop, and the possibility of publishing articles in chemical journals provided significant incentives to apothecaries to extend and deepen their chemical investigations. By way of two examples, the professional careers of Andreas Sigismund Marggraf and of Martin Heinrich Klaproth, part four will summarize collective conditions in eighteenth-century Germany that enabled apprenticed apothecaries to become renowned chemists.

PHARMACEUTICAL APPRENTICESHIP

Pharmacy was an old art or craft, one which had led to the establishment of apothecary's shops in Europe in the late Middle Ages. Chemical operations such as distillation and decoction were performed in apothecary's shops long before "chemical remedies" were introduced during the seventeenth century as a consequence of the iatrochemical movement in the vein of Paracelsus and his followers. Hence the pharmaceutical handicraft was not unprepared for the chemical innovations that were spurred by Paracelsian physicians and chymists.[15] We know from medical edicts, pharmacopoeias and other types of pharmaceutical books published in eighteenth-century Germany, as well as from *Arzeneitaxen* and inventories of apothecary's shops, that chemical remedies were broadly accepted around 1700.[16] The Brandenburg medical edict from 1698, for example, ordered that the "chemical remedies" (*chimische Medicamenta*) must not be purchased from "vagrants and laborants" but prepared and sold by apothecaries in their own "laboratories."[17] Especially the work of the historians of pharmacy Wolfgang Schneider and Erika Hickel provided ample evidence, based on their detailed analysis of German pharmacopeias and experimental reconstruction of their recipes, that the chemical remedies were an accepted part of the *materia medica* in early eighteenth-century Germany.[18] By this time the term "chemical remedies" was as ubiquitous as the division of remedies into *simplicia* and *composita*, and the further division of the latter into Galenic *composita* and "chemical preparations" or "chemical remedies." Nevertheless, the manufacture of chemical remedies meant the introduction of many new materials, reagents, and sequences of techniques into an existing artisanal tradition, and the way German apothecaries rose to this challenge and integrated chemical manufacture into their art and their system of apprenticeship is not yet sufficiently understood.

The pharmaceutical apprenticeship in eighteenth-century Germany normally took five to six years, followed by six to eight years of service as a journeyman.[19] In most German states the prospective apothecaries had to pass an examination before the local medical board (*collegium medicum*) and to swear an oath to the governmental pharmaceutical or medical edict (*Apothekerordnung, Medizinalordnung*). Unlike in France and Italy, in most German states apothecaries were not organized in guilds, and all regulations of the pharmaceutical apprenticeship and of the rights and duties of apothecaries were organized by the governments. In the seventeenth and early eighteenth centuries the German pharmaceutical and medical edicts fixed only a few, very general rules for apprenticeship. Apprentices were accepted at the age of 14 or 15 after schooling, with primary school education (*Volksschule*) considered sufficient if complemented by courses in Latin. They were trained in the *officine* and the laboratory by a master apothecary (*Principal*), either the owner of the apothecary or his or her (in case of widows) administrator (*Provisor*), and by the older journeymen.[20] In some

German states, such as Prussia, the physicians of the *collegium medicum* were not only the final examiners but were also involved in the admission of apprentices. The Prussian medical edict of 1693 stipulated that apprentices be introduced to the *collegium medicum* or the town physicians at the beginning of their apprenticeship, and on this occasion promise to obey all regulations of the medical edict.[21] Apothecaries were asked to accept only apprentices who were docile and knew some Latin. The edict further regulated that apprentices had to clean utensils, take care of the fire and light in the laboratory and the *officine*, help the clerks dry herbal drugs, prepare and sell medicines, and to pay special attention to the measures, weights, and prices of the medicines. Apprentices were not allowed to prepare medicines for sale on their own before they had completed their fourth year of apprenticeship.[22] Journeymen were warned to be particularly careful and accurate when they prepared *opiata* and *purgantia*.[23] No further regulations were specified concerning the concrete practice of the manufacture of remedies and the content of pharmaceutical training. Instead, the edict pronounced a couple of rules concerning the moral virtues of the apprentices and journeymen. They had to pay respect to their master, contribute to their welfare, say their prayers, and abstain from drinking, gambling, and idleness.

The Prussian Medical-Surgical College

From the medical and pharmaceutical edicts, as well as the official pharmacopoeias and *Arzeneitaxen*, we get the picture that chemical remedies and laboratories for their manufacture were quite common in Germany around 1700, and hence that some chemistry must have been taught in the pharmaceutical apprenticeship. But as the more concrete practice and content of the pharmaceutical apprenticeship was not formally regulated, it varied considerably in the different German states and individual apothecary's shops, especially with respect to chemical training and education.[24] The Prussian state was among the first to undertake efforts to regulate the chemical (and botanical) education of apothecaries. As early as 1725, King Frederick Wilhelm I enacted a new medical edict ordering that all apothecaries who wanted to establish themselves in larger and smaller cities had to pass an examination in practical chemistry at the newly founded Medical-Surgical College (*Collegium medico-chirurgicum*).[25] The prospective apothecaries were required to have served at least seven years as journeymen, and to demonstrate their chemical and pharmaceutical knowledge and skill in public by performing several practical *processus pharmaceutico-chymici* in the classes of the new school's professor of practical chemistry;[26] afterwards there was an examination by the physicians (*medici*) of the *Ober-collegium-medicum*.[27] Pharmaceutical journeymen and apothecaries were also invited to attend the new school's chemical and botanical courses, most of which took place in the anatomical theater of the Berlin Society of Sciences (after 1744 called the Berlin Academy of Sciences and Fine Literature).[28] The first courses on chemical theory related to the preparation of chemical remedies (*chymia rationalis pharmaceutica*) were offered by the chemist Johann Heinrich Pott (1692–1777), and practical chemical courses were given by the professor of practical chemistry (*chymia practica*) Caspar Neumann (1683–1737), who was also Royal Court Apothecary. As we know from Neumann's lectures, his courses, held in the nearby laboratory of the Royal *Hofapotheke* (Court Apothecary's shop), situated in a side building of the city castle, concentrated on the preparation of chemical remedies and their practical uses.[29] The surviving list of matriculation to the Medical-Surgical School from 1730 to 1768 reveals that the number of its *studiosi pharmaciae* was small, namely, between two and nine every year.[30] Hence, the impact of this new school for pharmaceutical apprenticeship in Germany was limited, although it may have spurred changes elsewhere. For example, the University of Göttingen, founded in 1737, offered courses in pharmacy shortly after its foundation,[31] and in the last decades of the century several chemical–pharmaceutical boarding schools were established that educated prospective apothecaries alongside chemical manufacturers and other students (see below).

However, throughout the eighteenth century the majority of German apothecaries attended courses at neither universities nor professional schools; their apprenticeship continued to rely entirely on the master apothecary in the local apothecary's shop. How was the manufacture of chemical remedies taught in the traditional system of apprenticeship? How much chemistry and which parts of chemistry were included? From pharmaceutical textbooks and manuals as well as the official pharmacopoeias, *Arzeneitaxen*, and medical edicts, we learn that the manufacture of chemical remedies was indeed accepted in the early eighteenth century and even desired by German governments and the *collegia medica*. But they do not provide sufficient information about the local practices of apprenticeship, and the distribution of chemical manufacture of remedies between cities and provinces, let alone single apothecary's shops. Insight into these latter issues can be gained from letters by apothecaries, autobiographical notes, and the debates on pharmaceutical apprenticeship in the late eighteenth century.

Pharmaceutical Apprenticeship in Practice

In his autobiographical notes, the apothecary and chemist Johann Christian Wiegleb (1732–1800) described his six years of apprenticeship (1748–54) at C. F. Sartorius's *Marien-Apotheke* in the city of Dresden as a period of disappointment and endless torture.[32] Sartorius's apothecary's shop was administered by an old administrator (*Provisor*), and the teaching of the four apprentices was left entirely to the two journeymen, who, according to Wiegleb, treated the four boys "like slaves."[33] To learn, Wiegleb complained, meant nothing other than "drill" (*abrichten*), and "in the whole period [of six years] I was never provided with the reason and conception [*Grund und Begriff*] behind any single task."[34] Hence, Wiegleb continued, "I had to educate myself." He did so by reading the books he found in the small library belonging to Sartorius's apothecary's shop. From the books he then mentioned – Johann Helfrich Jüngken's *Lexicon chymico-pharmaceuticum* (1693), Oswald Croll's *Basilica chymica* (1609), Jean Beguin's *Tyrocinium chymicum* (1610), and Van Helmont's writings – it becomes clear what Wiegleb had in mind when he spoke of the "reason and conception" underlying pharmaceutical techniques: chemical knowledge.[35] But not even these books fulfilled what they promised, Wiegleb observed. They rather directed him to the "wrong way" of alchemy, as he realized some ten years later, when he was already the owner of an apothecary's shop in his father's town of Langensalza and read the more useful "newer" chemistry books such as Rudolph August Vogel's *Institutiones chemiae* (1755), which he later translated into German (*Lehrsätze der Chemie*, 1775),[36] Christian Ehregott Gellert's book on metallurgical chemistry (*Anfangsgründe zur metallurgischen Chymie*, 1751), the recent writings on plant chemistry by Johann Friedrich Cartheuser (professor of chemistry, pharmacy, and *materia medica* at the University of Frankfurt/Oder), as well as the books of the famous French chemist Pierre Joseph Macquer.

Wiegleb's autobiographical notes are, of course, more or less selective reminiscences of an apothecary who was a renowned chemist. But from his remarks we get a sense of how insecure education in chemistry was within the traditional system of pharmaceutical apprenticeship. Even during his stay as a journeyman at the Court Apothecary's shop (*Hofapotheke*) in Quedlinburg he "learned nothing."[37] On the other hand, he obviously did learn chemical techniques, for otherwise he would not have read books on chemistry to illuminate "the reason and conception [*Grund und Begriff*]" underlying the techniques. Books on chemistry and a growing book market were important social means to learn chemistry in the eighteenth century, as we know from other apothecaries' testimonies as well. The fact that Wiegleb found a small library in Sartorius' apothecary's shop, which contained books on alchemy and chemistry, was not unusual either. Eighteenth-century German apothecaries were artisans and merchants who manufactured, bought and sold commodities and at the same time often cultivated learned knowledge. Especially the richer owners of large apothecary's shops, situated in towns, equipped their *officines* with precious ceramic vessels, established mineral and botanical cabinets, and increased their prestige by setting up a library. Although the library of Sartorius obviously did not satisfy the expectations of a man like Wiegleb, when viewed from the perspective of many other arts and crafts at the time its mere existence must be highlighted as an important social condition for education and higher learning.[38]

The distinctive social and cultural status of eighteenth-century German apothecaries was also reinforced by their relationship with physicians. Much has been written about struggles for power and authority between apothecaries and physicans, especially physicians who were members of the *collegia medica* that controlled apothecaries. But the need for professional communication often spurred friendship between town physicians and apothecaries. In case of Wiegleb, his friendship with the Prussian Field Physician E. G. Baldinger, who established himself as a practical physician in Langensalza in 1764, meant encouragement to read more and newer books on chemistry. In his autobiographical notes Wiegleb even described his meeting with Baldinger as the turning point in his professional life, as it drove him away from alchemy and put him "on the right track to science."[39] But we should bear in mind that books alone and conversations with a physician, even one most knowledgeable in chemistry, would not have fallen on fertile ground had Wiegleb not learned chemical techniques and owned a laboratory that he could use both for the manufacture of remedies and to perform chemical experiments. In 1779, Wiegleb further extended the use of his laboratory when he established a private boarding school (*Privat-Institut*) for the "systematic teaching" of chemistry.[40] Until 1798, when his health declined, he taught some forty students in his school, an average of two every year. The students lived and took their meals in his house, used his library and participated in the chemical and pharmaceutical work he did in his laboratory. Among his students were well-known apothecary-chemists such as Sigismund Friedrich Hermbstädt (1760–1833) and Johann Friedrich A. Göttling (1753–1809) as well as the Birmingham manufacturer's son M. R. Boulton, to whom Wiegleb tried selling his method of indigo dyeing.[41] As his friend Stoeller pointed out, Wiegleb fielded requests from fathers living in diverse European countries "to accept their sons as students of pharmacy and chemistry, provide them with deeper insight into the natural science [*Naturwissenschaft*], or to teach them bourgeois economy [*bürgerliche Geschäfte*]."[42] It is in this context that Wiegleb wrote the two volumes of his famous chemistry textbook *Handbuch der allgemeinen Chemie* (1781), whose context went far beyond the chemical knowledge that is useful for pharmacy.

By the end of the eighteenth century, the conviction grew among apothecaries that chemical–pharmaceutical boarding schools were a better alternative to the traditional system of pharmaceutical apprenticeship, as the next example shows while also providing further insight onto the system of apprenticeship during the 1760s.[43] In an anonymous report of his apprenticeship in the 1760s, published in Johann Friedrich A. Göttling's chemical–pharmaceutical periodical *Almanach* of 1793, an apothecary living in a small country town suggested replacing the traditional system of apprenticeship with public pharmaceutical schools; the suggestion was strongly supported in a final remark by the journal's editor Göttling.[44] Nevertheless, this anonymous report presented a much more positive picture of the traditional pharmaceutical apprenticeship during the 1760s than Wiegleb's, despite the fact that the apprenticeship took place in a "mid-sized country town" rather than a city like Dresden:

> I was not so unlucky to get apprenticed by a man who was totally ignorant and lacked education, as so many are; rather my master, who lived in a mid-sized country town but had become acquainted with no apothecary's shop other than the one in which he was apprenticed and his own, loved the sciences, read a lot, and possessed a nice little library; at least he owned the newest chemical writings, inasmuch as chemistry was advanced at this time. He possessed several very good physical instruments, a small, but for the time very instructive, collection of minerals and an impressive collection of stuffed birds. ... He prepared true white magnesia [*Magnesia alba, Magnesium carbonicum*] from bitter salt [*Sal amarum, Magnesium sulfuricum*] by precipitation with potash; likewise he prepared milk of sulfur [*Lac sulfuris; Sulphur praecipitatum*] and several other things in a genuine way following good chemical principles. When he prepared extracts he was very careful, paying attention that the extract did not remain too long in copper vessels, which I unfortunately experienced very often in *officines* [*Officinen*]. In addition his wife was present in the house, caring for the discipline of the pupils. An older pupil who had almost finished his apprenticeship and myself did the work in the officine by ourselves, but the wife was always there keeping house and urging us to work hard. ... The older pupil was the dispenser and did the few tasks in the laboratory, where the master was also present. He [the

> older apprentice] knew exactly how to get rid of the hard and unpleasant work, so that I had to do all the work that a maid or servant should have done. To this belonged in particular the cleaning of vessels. ... After three years, when the older pupil had finished his apprenticeship, he left, and I replaced him as a dispenser. I knew little about the work in the laboratory, as I never had the opportunity to learn it, but fortunately there was a book of recipes for the preparation of the usual goods so that I learned quickly. Now I also began to learn the scientific part of our art by reading Gottlieb's *Neu eröffnete Apothekerschule* (1701), Helwig's *Apothekerschatz* (1709) and Kräutermann's *Wohlerfahrener Apotheker oder Anleitung zur Apothekerkunst* (1730), which were kept in the apothecary's shop for the older pupils along with the dispensatories of Württemberg and of Brandenburg.[45]

In this case, the master apothecary possessed the newest chemical writings, a collection of *naturalia*, and a well-equipped laboratory. He was also a skilled manufacturer of chemical remedies, whose main shortcomings apparently were indifference to his apprentices and an insufferable wife who replaced him in the training of apprentices.

Let us return to Quedlinburg, where Wiegleb had stayed in vain for one year in the *Hofapotheke*, having "learned nothing." Quedlinburg, a lively city of eight thousand inhabitants in the Harz region, was also the site of apprenticeship of two other German apothecaries – one less known, Johann Christian F. Liphardt (1758 or 59–1805), and one very well known, Martin Heinrich Klaproth (1743–1817). Both Liphardt and Klaproth were apprenticed in the *Ratsapotheke* of Quedlinburg, established in 1578. *Ratsapotheken*, which were owned by towns, were generally very well equipped. In case of Quedlinburg's *Ratsapotheke*, which still exists today, we have historical documents that show that the *officine* was located in the middle of the ground floor of the half-timbered front building and the large laboratory in a room of the side corridor; a room for materials (*Materialstube*) and another one for confectionery (*Zuckerkammer*) were on the first floor; two rooms for storing herbal drugs and for glass were in the attic; and two rooms for storing distillates, wine, brandy and aqua vitae in the cellar (see Figures 1–3).[46] As Quedlinburg was part of a Prussian protectorate, the apothecary trade was regulated by the Brandenburg-Prussian medical edict and pharmacopoeia, the *Dispensatorium Regium et Electorale Borusso-Brandenburgicum* of 1731. Like the Brandenburg medical edicts of 1685 and 1693 (see above), the subsequent Prussian medical edict of 1725 permitted only apothecaries to produce chemical remedies, privileged secret remedies excepted. The pharmacopoeia of 1731, which classified medicines alphabetically (in Latin) and presented short recipes for their preparation, included many newly introduced chemical remedies,[47] such as *clyssus antimonii, crocus Veneris, liquor martialis, magisterium Saturni, spiritus nitri dulcis* and *spiritus salis dulcis* (which contained substances later called "ethers"), *tinctura antimonii* and so on. All chemical remedies were prepared in a "laboratory," equipped with furnaces, a chimney, alembics and other distillation apparatus, instruments for melting and cupellation, and various kinds of vessels for dissolution, filtration, and precipitation.[48] The laboratory of the Quedlingburg *Ratsapotheke* was well suited for chemical work, as it had massive stone walls, a stone floor, and cross vaulting that provided fire protection (see Figure 2).

Figure 1. Photograph of the *(Adler- und) Ratsapotheke* of Quedlinburg on Kornmarkt 8 today, by the author

Figure 2. Photograph of the ancient room (with its vaulted ceiling) used as the laboratory of the *(Adler- und) Ratsapotheke* of Quedlinburg, by the author

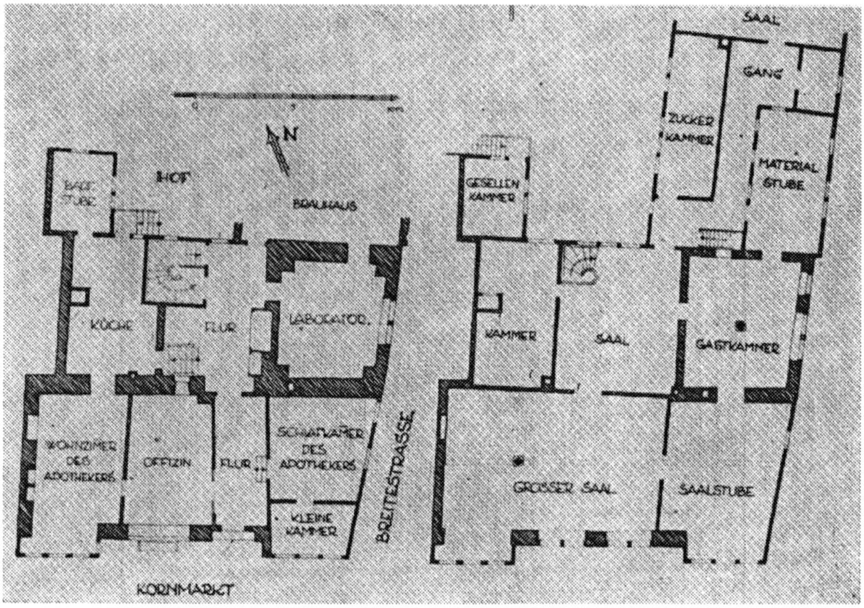

Figure 3. Seventeenth-century ground plan of the Quedlinburg *(Adler- und) Ratsapotheke*, reconstructed by Gründhagen, *Apotheken in alter und neuer Zeit*, 31

Thus we can take it for granted that the *Ratsapotheke* of Quedlinburg was a place where the manufacture of chemical remedies was thriving.[49] But the pharmaceutical apprenticeship seems to have been disappointing, especially for Klaproth, who began his apprenticeship there in 1759. Klaproth recalled: "I cannot pride myself in having received lectures from my master. Rather, as was customary at that time, I had to be satisfied with observing my older fellows doing their handicraft and with the occasional reading of one or two antiquated apothecary books."[50] Liphardt, whose apprenticeship in the *Ratsapotheke* of Quedlinburg started in 1773, experienced better treatment. From 1763 until 1785, the *Ratsapotheke* was run by Viktor Friedrich Bollmann (1712–89), who was born into a family of physicians living in Quedlinburg.[51] As a journeyman Bollmann had learned pharmacy in six different towns, which presumably made him acquainted with a broad range of pharmaceutical techniques; and, as the son of a physician who had alchemical interests, he also knew some chemistry. In a letter to J. F. A. Göttling Liphardt wrote some ten years later that he was lucky to be apprenticed to a man like Bollman, although at the time he always believed that there must be other places where one could learn more. Hence, he remarked, "I was waiting with impatience for the time when I could work in a second laboratory."[52] But what he experienced elsewhere taught him that he was wrong. In his first year as a journeyman he learned quickly that his "stock of science" was entirely useless for the kind of work he did.[53] Liphardt went on to complain that many apothecaries took apprentices for the sole aim of having them do the hardest and dirtiest work in the shop. He pointed out, in particular, that in many *officines* the apothecary "obtained all goods whose manufacture required chymical knowledge and instruments from larger places; hence, the apprentice acquired only knowledge about the external form of things rather than about their components and ways of preparation."[54] He then suggested that "apothecaries in small towns, which cannot subsist without a grocery (*Materialhandel*) should have more scruples about taking apprentices."[55] On the other hand, he also conceded that the hope of many parents that their sons would get a better education in a large apothecary's shop was not always fulfilled, as the more famous apothecaries often belonged to "those learned men [*Gelehrte*], who do not like to communicate their skills and instead work for their own benefit." "It is well known that mere theory without accumulating experience on one's own is useless," he continued.[56] In addition, the smaller apothecary's shops had the advantage that apprentices learned to be more economical with their goods. He finally recommended that anybody who had learned in a large "chymical laboratory" should also spend a year in a small apothecary's shop.[57]

Liphardt's detailed report of his apprenticeship confirms Wiegleb's autobiographical reminiscences in at least two ways. First, pharmaceutical apprenticeship resembled apprenticeship in other contemporary arts and crafts inasmuch as it involved a lot of dirty, repetitive handiwork. But second, it also often included the learning of chemical techniques in the laboratory, and books on chemistry and pharmacy were often on hand for improving chemical knowledge. With respect to learning chemical techniques, Liphardt also made another observation: despite the fact that medical and pharmaceutical edicts required apothecaries to prepare the chemical remedies by themselves, there were also apothecaries, especially owners of small apothecary's shops in the countryside, who bought chemical remedies from grocers (*Materialisten*, later druggists), travelling laborants, distillers, and confectioners.[58] In such cases we may assume that the apothecary's shop did not include a well equipped laboratory. We have no exact data about the proliferation and equipment of chemical–pharmaceutical laboratories in eighteenth-century Germany. From the fact that they were taken for granted in medical edicts and pharmacopoeias and often mentioned in letters and published reports by apothecaries – and further given the many remaining chemical–pharmaceutical instruments from that time in today's pharmaceutical museums (see Figure 4) – we may conclude that they were indeed very widespread. But with respect to the questions studied in this paper we must bear in mind that there were local differences and that by no means all eighteenth-century German apothecaries were interested in chemistry or had the opportunity to learn chemical techniques that went beyond distillation and decoction. As a heuristic device we may distinguish four different groups of eighteenth-century German apothecaries: first, apothecaries who did not possess a well equipped laboratory and rarely prepared chemical remedies themselves; this group was presumably a small minority. Second, apothecaries who prepared a considerable range of chemical remedies themselves, but whose interests in chemistry never went so far that they read the newest chemical publications such as chemical periodicals. Third, apothecaries who prepared a large range of chemical remedies, read chemical periodicals, occasionally published articles in chemical journals,

Figure 4. Chemical–pharmaceutical instruments (17th–19th century), courtesy of the Deutsches Apotheken-Museum, Heidelberg

but did not become well known as chemists. The third and second groups constituted the bulk of eighteenth-century German apothecaries; however, as we will see below, they were distributed extremely unequally, the third group being concentrated in larger cities. Finally, there was the much smaller fourth group of apothecaries who became renowned chemists in eighteenth-century Germany.

It is not surprising that apothecaries who belonged to the fourth group were especially critical of the traditional system of pharmaceutical apprenticeship, established private pharmaceutical boarding schools, and aimed at a more comprehensive "scientific education" for apothecaries. When these apothecary-chemists complained that their own apprenticeship lacked proper teaching in chemistry they presumably had in mind a broad range of forms of chemical knowledge, practical and theoretical, which went beyond pharmaceutical usefulness.[59] It was different with apothecaries who belonged to our middle field, as the example of Ernst Wilhelm Martius (1756–1849) shows.[60] E. W. Martius, who was Court Apothecary in Erlangen from 1792 and lecturer for pharmacy and pharmaceutical *Warenkunde* (knowledge of commodities) at the University of Erlangen after the Napoleonic wars, was apprenticed in the 1770s in Erlangen's *Hofapotheke*, which was run by his uncle. In his autobiography he described his apprenticeship as follows:

> At first my service required only mechanical work such as pulverizing and cutting roots. ... The scientific part of the business I learned less through personal teaching than through careful observation and paying attention to the handiwork and practices of the other workers. In order to know more about chemical processes, I had to read the contemporary textbooks such as Gleditsch, *Verzeichnis der gewöhnlichsten Arzeneigewächse* (1769), Lemery's *Dictionnaire der Drogen*, Spielmann's *Anleitung zur Kenntnis der Arzeneimittel* (1775), Lösecke's *Materia medica* (the sixth edition of which was edited by Gmelin), and, above all, the *Dispensatorium Würtembergicum*, which had the greatest reputation as a practical pharmaceutical code. ... By the way, the latter also contained improved chemical recipes, and therefore I am well aware that this valuable book was more useful to me than many other pharmaceutical textbooks and manuals we possess today. The first edition of Hagen's *Lehrbuch der Apothekerkunst*, which was an extremely useful manual at the time, did not appear until 1778. My uncle's chemical studies relied on the writings of Bergius, Linné, Boerhaave, Stahl, and Teichmeyer. ... At that time an apprentice was very busy getting well acquainted with the German and Latin nomenclature of the numerous raw materials and compound remedies. For this purpose I received an *Arzeneitaxe*. In order to became trained in the business of the *receptarius* I had to help him do his work.[61]

Like Wiegleb, Klaproth, and Liphardt, Martius stated that he missed personal teaching in his apprenticeship, and instead learned by doing and observing. But he complained much less about the lack of chemical education than the three apothecary-chemists. As late as 1847, at the age of 91 and at a time when many prospective German pharmacists also studied pharmacy at universities, he had no qualms about restricting chemical knowledge to its pharmaceutically useful parts as presented in pharmaceutical books.

To round out our picture of pharmaceutical apprenticeship, I will describe a sixth example that shows that there were also apothecary's shops in eighteenth-century Germany where apprentices received personal training and advanced chemical education. In a letter of 1787 an apothecary who remained anonymous described to Göttling his own attempts to improve pharmaceutical apprenticeship.[62] His novel method of "scientific education," he stated, would be much shorter than the usual six-year period, namely four years, not least since he accepted apprentices only at the age of 16, after they had received a good school education. In the first year, the apprentice had to learn the names of the numerous raw materials (*simplicia*), Galenic *composita*, and chemical preparations as well as measuring and weighing; this was done in addition to the "usual business of an apprentice," that is, presumably cleaning, errands and the like. For the former purpose, the apprentice had to read on his own a Latin apothecary book, which also served as Latin practice. Once a week he also had to write a letter, and in addition he received a copy of Bindheim's *Rapsodien der philosophischen Pharmacologie*, a book that the author strongly recommended to every young apothecary. Not until his second year did the apprentice get to read a "chemical–pharmaceutical book in the proper sense," namely Karl Gottfried Hagen's *Lehrbuch der Apothekerkunst*, of which the author recommended especially the newly published third edition of 1786 (see Figure 5).

Hagen (1749–1829), who was Court Apothecary in Königsberg and professor of medicine at Königsberg University, regarded pharmacy as both an art and a science, which was a "part of chemistry." As pharmacy "included almost all kinds of operations that existed in the other parts of the entire field of chemistry as well, it was an epitome of chemistry; and nobody could achieve perfection in chemistry without first learning all pharmaceutical operations, in theory and practice."[63] His *Lehrbuch der Apothekerkunst*, which became the most famous pharmaceutical textbook around 1800, was divided into three parts. Part one dealt with the art and science of pharmacy in general, including descriptions of instruments and units of measurement, and moral instructions to the apothecary. In these instructions he reminded the apothecary that he must never rely on *Materialisten* and other merchants but rather prepare the chemical remedies in his own laboratory.[64] He further demanded that the master apothecary instructs his apprentices about the way to identify remedies and about their "rational preparation," and further that the journeyman should not work exclusively in the *officine* as a dispenser (or *Rezeptarius*, a person who weighed simple ingredients and mixed them together to prepare Galenic compound remedies) or exclusively in the laboratory, but rather do both kinds of work.[65] Part two was concerned with the "raw medicines" (*simplicia*), divided according to the three natural kingdoms, their collection and their preservation. The third part about "the pharmaceutical operations" described in a general way all kinds of mechanical and chemical operations, and the fourth part presented classes of "pharmaceutical preparations."

Our master apothecary carefully followed Hagen's advice, not only by handing out his textbook to the apprentices but also by administering weekly examinations

Lehrbuch
der
Apothekerkunst

von

Karl Gottfried Hagen

der Arzeneigelahrtheit Doktor und Professor,
Königl. Preußischem Hofapotheker, der Röm. Kaiserl.
Akademie der Naturforscher und der Berlinischen
naturforschenden Gesellschaft Mitglied.

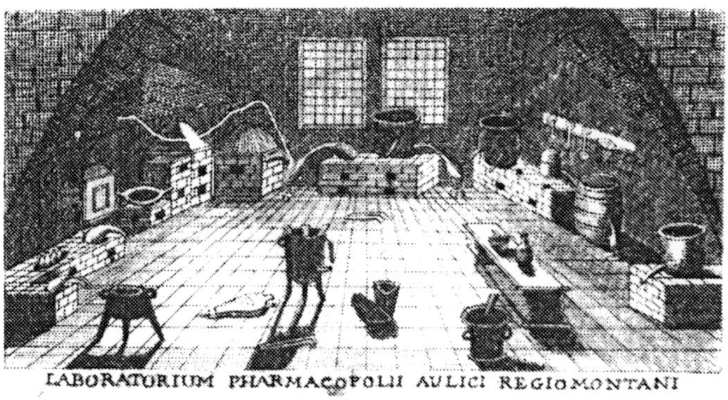

Dritte, rechtmäßige und verbesserte Ausgabe.

Königsberg und Leipzig,
bey Gottlieb Lebrecht Hartung,
1786.

Figure 5. Title-page of Hagen's *Lehrbuch der Apothekerkunst*, depicting the laboratory of the *Hofapotheke* in Königsberg; courtesy of the Universitätsbibilothek Göttingen

on the reading material. With respect to the first part of the book (the one on natural raw materials) he mentioned that he also instructed his apprentices in the natural history and provenance of materials. Referring to the "pharmaceutical-chemical part" of the textbook, he continued: "I acquaint the apprentice with the properties of the bodies, especially the salts, and with their compounds; but first of all I try to teach them the idea of the chemical affinities [*Verwandtschaft*] of bodies." He added that practical instruction was crucial too, and that he never failed to highlight the "chemical principles" upon which the operations were based. Another element of instruction was botany, in which he aimed at "systematic knowledge of plants" based on the Linnean system.[66] After this systematic teaching in the first two years, it was expected that the apprentices had progressed sufficiently to occupy themselves with "more chemical writings" such as Wiegleb's *Handbuch der allgemeinen Chemie* (1781). In the foreground of the teaching in the third year was mineralogy, including instruction on the master apothecary's own mineral cabinet, botanical exercises in summer, and *materia medica*. In the fourth year chemical instruction was crowned by the reading of Crell's brand new chemical periodical, the *Chemische Annalen*. Our apothecary did not even "fear the expense of repeating all ambiguous experiments [*Versuche*], whereby he [the apprentice] was instructed to perform experiments himself in the future."[67] Further reading of books such as Blumenbach's *Handbuch der Naturgeschichte* (1779), Kirwan's *Anfangsgründe der Mineralogie* (1785), Erxleben's *Anfangsgründe der Naturlehre* (1785), the German translation of Macquer's *Chemisches Wörterbuch* (1781), and Weigel's *Grundriß der reinen und angewandten Chemie* (1779) completed the pharmaceutical education. Although this last case may be an extreme example, we get a sense of the diversity of the local conditions of pharmaceutical apprenticeship in Germany in the late eighteenth century.

Boarding Schools

In reaction to the diversity and instability of pharmaceutical apprenticeship in eighteenth-century Germany, by the end of the century a reform movement developed, for which Göttling's *Alamanch* and Trommsdorff's *Journal der Pharmacie* became a public forum.[68] Suggestions for reform largely agreed in three aspects: the request for closer regulation of pharmaceutical education, the establishment of professional schools, and the idea of transforming the traditional apprenticeship into "scientific" education by teaching chemistry and botany. In two anonymous articles published in Göttling's *Almanach* in 1793, one apothecary argued that a good apothecary needed to be familiar with botany and with "all chemical–pharmaceutical preparations, both theoretically and practically." In order to guarantee this, he suggested printed curricula issued by the governments; the other apothecary proposed public schools.[69] In the same issue of the *Almanach*, the young Erfurt apothecary Johann Bartholomäus Trommsdorff (1770–1837) pleaded for "scientific pharmacy," which included the acquisition of knowledge about natural history and "applied chemistry."[70] Shortly thereafter Trommsdorff edited a journal of his own, entitled *Journal der Pharmacie für Aerzte, Apotheker und Chemisten* ("Journal of

Pharmacy for Physicians, Apothecaries, and Chemists"), whose main goal was to "extend the scientific study of pharmacy."[71] In the first issue of this journal he published a paper on his own methods of pharmaceutical apprenticeship, which stressed the systematic teaching of chemistry.[72] Trommsdorff's teaching of chemistry started with "its pharmaceutical part." The apprentice read useful books and was instructed by Trommsdorff in the laboratory, reporting his observations in a notebook that was reviewed by Trommsdorff every week. He was then made acquainted step by step "with the entire range of chemistry," by reading first Göttling's *Almanach* and then Crell's *Chemische Annalen* as well as textbooks on chemistry by Gren, Hermbstädt, Westrumb and Wiegleb. By the end of the four years of apprenticeship, the apprentice had to make his own chemical preparations in the laboratory. A year later, prospective apothecaries also had the possibility of training in Trommsdorff's *chemisch-physikalische und pharmaceutische Pensionsanstalt für Jünglinge* (chemical–physical and pharmaceutical boarding school for boys) in Erfurt.[73] In addition to this latter boarding school in Erfurt (1795–1828) and Wiegleb's in Langensalza (1779–98), there was the renowned *chemische Pensionsanstalt* (chemical boarding school) of Sigismund Friedrich Hermbstädt in Berlin that had opened in 1789. Hermbstädt's school addressed "prospective apothecaries or other curious (*wißbegierig*) boys who wanted to become chemists."[74] The students were taught "chemistry in its entirety" as well as neighboring disciplines such as physics (*Naturlehre*), mineralogy, pharmacy, *materia medica*, and practical, chemical analysis, the latter including assaying and metallurgical chemistry. Students performed their analyses in "a laboratory of their own, equipped with the necessary instruments and materials."[75]

Artisanal Apprenticeship Combined with Higher Education

With respect to our original question concerning the way eighteenth-century German apothecaries learned chemistry during their apprenticeship, the following four aspects can be summed up. First, pharmaceutical apprenticeship, as a rule, took place in the *officine* and laboratory of an apothecary's shop. The apprentice learned by observing the work of the master and journeymen and by doing. In accordance with the habits of apprenticeship in the eighteenth-century arts and crafts more broadly, they had to do much repetitive handiwork. Second, as chemical remedies were broadly accepted at the time, and even strongly supported by the German governments, the majority of apothecaries owned a chemical–pharmaceutical laboratory for the manufacture of chemical remedies. Hence, during their six years of apprenticeship most pharmaceutical apprentices became acquainted with chemical instruments and chemical techniques. Learning chemical techniques relied mainly on the observation of masters and journeymen and practical repetition. Furthermore pharmaceutical apprentices acquired connoisseurship about thousands of materials – raw materials of vegetable, animal, and mineral origins, Galenic compositions, chemical reagents and chemically prepared remedies – as well as skills in weighing and measuring. They also read books on pharmacy and chemistry as well as *Arzeneitaxen* and pharmacopoeias, written in Latin. Hence, their apprenticeship exhibited characteristics of both artisanal apprenticeship and higher education. With respect to the latter, however, there was, third, great diversity among local apothecary's shops. One extreme was personal training along with "scientific education" that ranged from Linnean taxonomy to laws of chemical affinity, while the other extreme was the restriction to a small range of pharmaceutically useful chemical preparations and texts. Fourth, pharmaceutical apprenticeship was in a state of dynamic transition in eighteenth-century Germany. The latter fact becomes manifest not only in the diversity of pharmaceutical training but also in suggestions for reform as well as actual innovations, governmental and private, such as the establishment of chemical examinations of prospective apothecaries at the *Collegium medico-chirurgicum* in Berlin and of private chemical and pharmaceutical boarding schools later in the century.

THE ROLE OF TRAVEL

Historians of technology and economics have pointed out the significance of travel and migration of journeymen for innovation and technological change in the arts and crafts in early modern Europe.[76] In the eighteenth century, pharmacy was a rapidly transforming art, due to the introduction of novel materials and chemical techniques for the manufacture of chemical remedies. This process of innovation in the pharmaceutical art received considerable stimulus from the system of service as journeymen and the exchange of knowledge and skill through traveling to many different sites of pharmaceutical and chemical manufacture, in Germany and abroad. Furthermore, traveling also provided prospective apothecaries with ample opportunity for higher learning and for extending their chemical knowledge beyond the boundaries of the pharmaceutically useful. As there was no formal regulation in Germany concerning journeymen's travel and the number of places they had to visit, there were great differences in this respect. For example, J. C. Wiegleb, whose professional career was described in some detail above, had little opportunity to travel during his time as a journeyman. His one-year stay as a journeyman in the *Hofapotheke* of Quedlinburg ended with his uncle's death, which forced him to return to Langensalza in order to adminster his uncle's apothecary's shop. In 1770 Wiegleb made good what he had missed on a tour of industrial sites in the Netherlands, where he wanted to learn more about the chemical techniques of "laborants."[77] A more typical example is M. H. Klaproth, who had five positions as a journeyman: in the *Ratsapotheke* of Quedlinburg, the *Hofapotheke* in Hanover, the apothecary's shop *Zum goldenen Engel* in Berlin, the *Ratsapotheke* in Danzig, and the apothecary's shop *Zum weißen Schwan* in Berlin (see below).

The autobiography of E. W. Martius sheds some light on the way eighteenth-century pharmaceutical journeymen extended and refined their knowledge by traveling. Martius traveled extensively – he served at ten different apothecary's shops before becoming Court Apothecary in Erlangen – and, as he never became a well-known chemist, he is not too exceptional a case. When Martius finished his apprenticeship in Erlangen in 1777, he first took a position at the *Hofapotheke* of Coburg. The Court Apothecary of Coburg was an alchemist, but Martius was obviously not very impressed by the latter's chemical expertise. The third station was the *Reichsstadt* Regensburg, which was renowned for its social and scientific life. In this "intellectually exciting environment," Martius remembered, his love for the "scientific art" of pharmacy flourished, especially through learning about the newest discoveries and inventions in chemistry.[78] Although he was quite reticent about the latter, during this time he obviously became acquainted with Göttling's new *Almanach*, the first issue of which was published in 1780 (see below). Furthermore, during his three-year stay the libraries of the abbey and the University of Regensburg offered ample opportunity for him to improve his botanical knowledge. In 1782 Martius nevertheless felt that he wanted to see more of the world, and so inquired about new positions with a merchant (*Materialist*) in Frankfurt (am Main), who was a sort of an agent for apothecaries and journeymen,[79] and then proceeded to Dillenburg, a small town in a mining region. There he not only learned more about medicine and physics through conversations with two excellent physicians, but also enjoyed botanical excursions with a female botanist, a "Demoiselle Katarina Helena Dörrienwith."[80] Furthermore his acquaintance with a local mining official (*Bergrat*) introduced him to mineralogy, mining, and metallurgy. The next town, the *Reichsstadt* Wezlar, was an equally stimulating place, where people conversed about the brand-new fashion of hydrogen balloons, and where he came in contact with Lorenz Crell, editor of the *Chemische Annalen* (see below). In 1785, Martius took a position in an apothecary's shop in the university city of Strasbourg, whose laboratory enabled him to "apply his improved chemical knowledge in practice."[81] The owner of this apothecary's shop was a well-educated man and "excellent chemist" named Hecht, who cared about teaching his apprentice and his three clerks, and gave them especially good advice in the laboratory.[82] In this apothecary's shop, Martius also met an alchemical celebrity of the time, Count Cagliostro, who had several of his secret remedies prepared there. The next station was in the university city of Mainz, where he obtained a position in the *Hofapotheke*. Two years later, his uncle informed him about the possibility of administering his apothecary's shop in Baiersdorf, a small town near Erlangen. This position was an important step in his career, and brought him the offer to become administrator (*Provisor*) of an apothecary's shop in Regensburg, where he stayed from 1788 until 1791. In 1791, he returned to Erlangen to become Court Apothecary, after passing an oral examination in practical chemistry conducted by the three physicians on the medical faculty of the University of Erlangen. The fact that he had improved his chemical, botanical, mineralogical and metallurgical knowledge in so many different places contributed significantly to his reputation and his being accepted as Court Apothecary.

The high esteem that travel enjoyed in eighteenth-century Germany for learning in all areas of chemistry is also apparent from the fact that patrons funded travel by promising apothecaries. For example, from September 1787 until February 1788 the Weimar apothecary Johann Friedrich A. Göttling, known as the publisher of the chemical–pharmaceutical periodical *Almanach*, traveled to England at the expense of Duke Carl August of Saxe-Weimar.[83] The journey was part of an arrangement between J. W. Goethe and the Duke, who wished to appoint Göttling to a new chair of chemistry and technology to be established at the University of Jena.[84] As Göttling had no university education, he first received a stipend to study at the University of Göttingen for two years and then to travel to England, as part of this higher education. His first stop was London, where he visited the laboratories of the physicians and chemists Bryan Higgins (1737 or 1741–1818) and George Pearson (1751–1828), and attended chemical lectures. In London, Göttling was full of admiration for the "many rich shops of merchants on the streets, illuminated in the evening, among which the many shops of druggists and chemists stand out."[85] Yet he also warned the Germans that most of the English who called themselves "chemists" knew "little or nothing about chemistry" and made their chemical preparations merely "in the manner of craftsmen" (*handwerksmäßig*).[86] The London Apothecary's Hall, however, with its two big laboratories, "where all chemical preparations are made in large quantities," received his unqualified praise.[87] From London his journey proceeded to Oxford and then to Birmingham, where he visited the factories of the chemist, geologist, and inventor James Keir (1735–1820). Another highlight was his visit to Joseph Priestley (1733–1804), who lived in a country house near Birmingham and had a "very well-equipped laboratory," including a "very good apparatus for the artificial airs."[88] Supported by Joseph Priestley and the industrialist and metallurgist Matthew Boulton (1728–1809), he also attended the monthly meetings of the Lunar Society. The last stop of his journey was Anglesey with its famous copper mines. Shortly after his return to Weimar, Göttling received a Ph.D. at Jena in recognition of his prior publications, and in 1789 he was appointed professor of chemistry and technology to the philosophy faculty there.

A similar professional career, in which travel played an even more important role, was that of Caspar Neumann (1683–1737) back in the early eighteenth century. Neumann was Court Apothecary from 1719 until his death in 1737, an ordinary member of the Berlin Society of Sciences from 1721, and, from 1723, the first professor of practical chemistry at the newly founded Medical-Surgical College (*Collegium medico-chirurgicum*) in Berlin.[89] His career began with an ordinary pharmaceutical apprenticeship in his hometown of Züllichau. In 1704 he went to Brandenburg and Berlin, where he served as a journeyman in one of Berlin's privileged apothecary's shops. Shortly afterwards he found a position in the laboratory of the *Hofapotheke* of King Frederick I, and from 1705 to 1711 he was a journeyman in the Royal Prussian Travelling Apothecary. During this period he accompanied King Frederick I on his travels to Holland and Karlsbad, and gained the King's favor along with an offer to study at a university or, alternatively, to travel "in order to learn the chemical art abroad."[90] As the second offer pleased him more, in 1712 Neumann traveled to the Harz region to visit mines and smelting works.[91] In the mining towns of Clausthal, Andreasberg, Zellerfeld and Goslar he learned metallurgy and assaying. Later in his chemical lectures, Neumann repeatedly illustrated his understanding of the chemical behavior of metals by referring to his observations of craftsmen's operations in smelting works.[92] On further travels through Germany he visited apothecary's shops, glassworks, foundries, private laboratories, medical and botanical gardens, and universities. His next stops were Amsterdam, Utrecht, and Leiden, where he met the famous physician and chemist Herman Boerhaave (1668–1738). He then visited London, but upon King Frederick's death in early 1713 he suddenly lost his financial support. An offer by the wealthy Dutch surgeon A. Cyprian to run his chemical laboratory in London enabled him to stay for the next three years and to improve his knowledge in chemistry. In 1716, Neumann joined the entourage of King George I for a trip to Hanover, which allowed him to resume his contacts with the Prussian government. Supported by Georg Ernst Stahl (1659–1734), who had been the First Royal Physician to the Prussian King since 1712, he reentered Prussian service. He again received financial support to travel to England, France, and Italy until a suitable position in the Royal *Hofapotheke* became available. In Paris he became acquainted with René-Antoine F. de Réaumur (1683–1757), and with the apothecary-chemists Etienne-François Geoffroy (1672–1731), his younger brother Claude-Joseph Geoffroy (1685–1752), and Simon Boulduc (1652–1729), the chemical demonstrator in the *Jardin du Roy*. It appears that he often visited Boulduc in the chemical laboratory of the *Jardin du Roy*, since he often criticized the latter's analyses of plants in his later lectures. By contrast, he always praised the chemical work of E. F. and C. J. Geoffroy, including E. F. Geoffroy's famous table of affinities. When Neumann returned to Berlin in 1719, via the Saxon mining region, he was immediately appointed Court Apothecary and furnished with the financial means to reconstruct and expand the laboratory of the *Hofapotheke*.[93]

THE ROLE OF JOURNALS

Traveling was a means of proliferating chemical knowledge and techniques in the eighteenth-century pharmaceutical art as well as a way to become a learned apothecary-chemist. As it was a habit of both journeymen and Enlightenment savants, it was a hybrid endeavor that matched apothecaries' mixed goals, commercial and epistemic. Travel was, of course, only one stimulus for learning chemistry. We now turn to additional institutions that spurred apothecaries to study chemistry and contributed to the emergence of apothecary-chemists. The existence of a book market and of books on chemistry has already been mentioned. Even more important was another new institution that provided opportunity not only for the receptive learning of chemistry but also for communication and active contribution to the development of chemical knowledge: chemical journals. Karl Hufbauer pointed out the significant role played by Lorenz Crell's chemical periodical in the emergence of the German chemical community. Hufbauer asserted that before Crell began to publish the *Chemische Annalen* in 1778, "for want of an adequate means of communication with one another, German chemists had not yet coalesced into a community." "Only a periodical," he added, "that could serve as a forum would establish the regular communications necessary to bind them into a German chemical community."[94] With a slightly stronger emphasis on the epistemic function of scientific journals, William Brock stated that "chemical periodicals are particularly important; from the late eighteenth century they came to replace the monograph for conveying new chemical knowledge and for settling controversial issues."[95] In order to function as a forum for communication and exchange of knowledge in the very heterogeneous group of eighteenth-century German chemists, Crell's journal had to abstain from too narrow a definition of erudition. Accordingly, in the preface to the first issue of his journal he emphasized that it would report each single observation, his own role being restricted to that of a "collector."[96] The journal, he asserted, was "completely open to each chemist, whether he is concerned with the processing of metals, the preparation of remedies, or with experiments for pleasure."[97] For this purpose it was important that all contributions be written in the vernacular, as in the case of the second chemical and pharmaceutical journal that followed Crell's in 1780, namely Göttling's *Almanach* (see below). The journal's goal was not only to proliferate "chemistry as a science," but also to "exert some influence on various aspects of everyday life."[98]

As we saw above in the introduction, Crell very successfully encouraged apothecaries to read his journal and to contribute to it. Almost 50% of the subscribers to his journal were apothecaries, and they were also its most active contributors. We may conclude from this fact that journals played a significant role not only in forming the German chemical community but also in forming the persona of apothecary-chemist. But how informative are the above numbers with respect to this latter issue? What share of the entire community of German apothecaries did the subscribers to Crell's journal represent? In order to answer this question we must correlate the number of subscribers to Crell's journal with the entire number of German apothecaries in the same period. Fortunately we have quite a good overview of the number of Prussian apothecaries at the time; it is based on a list which Sigismund Friedrich Hermbstädt put together for the Prussian king in 1798 in preparation for war with Napoleon and the corresponding need to recruit field apothecaries.[99]

According to this list, in 1798 there were 337 apothecaries in Prussia, including owners of apothecary's shops and administrators (*Provisors*). Their apothecary's shops were distributed over almost 300 towns, with the highest concentration in Berlin, which had twenty-four apothecary's shops. This large number of apothecary's shops may seem surprising, but we must bear in mind that at the time apothecaries sold not only remedies but also a range of other commodities such as coffee, tea, tobacco, spices, confectionery, pigments and tints, soap and cosmetics, wine and brandy. In the provinces, apothecary's shops survived mainly through this additional trade, as can be seen, for example, in complaints that the sale of beer, wine, and brandy transformed apothecary's shop into pubs. Presupposing that the number of apothecary's shops and apothecaries had not changed as rapidly as the entire population of Prussia, which almost doubled between 1784 and 1800, comparison of the total number of Prussian apothecaries with the number of Prussian subscribers to the *Chemische Annalen* yields the following result.[100] There were a total of 43 Prussian subscribers compared to 337 apothecaries, that is, 12.8%; if we assume that the number of privileged apothecary's shops increased slightly in the period between 1784 and 1798, in keeping to some extent with the enormous increase of the population, and assume that there was approximately a total of 300 apothecaries in the period between 1784 and 1789, the percentage of subscribers would be 14.3%. This proportion of subscribers was distributed extremely unequally. Twenty-five subscribing apothecaries (and ten apprentices) lived in Berlin, that is, practically all apothecaries in Berlin were subscribers to Crell's journal. Compared to Berlin, the percentage of subscribers among apothecaries in the rest of Prussia was low (approximately 6%).[101]

The insight that can be gained from these numbers is limited, not least since the numbers are not based on exact correlation. But two conclusions can be drawn. First, Berlin was the center of apothecaries' chemical activities in late eighteenth-century Prussia. Second, the community of apothecaries was heterogeneous with respect to their interests in chemistry, and only a small part of them had a distinctively strong interest in all areas of chemistry beyond the pharmaceutically useful.[102] But this part was by no means a negligible minority, as we must bear in mind that there was a second, competing periodical at the time, namely Göttling's *Almanach* (see Figure 6), which addressed apothecaries more directly. Furthermore, since Crell used the journal to improve his income, the journal was quite expensive.[103] In his autobiography, E. W. Martius recalled that during his stay in Wezlar (1783–85) Crell asked him to advertise his journal among apothecaries and to find new subscribers, which Martius did successfully. Yet he added that subscription to Crell's journal meant a "pecuniary sacrifice that apothecaries were not yet used to at the time."[104] Hence several of his new subscribers changed their minds shortly thereafter. This may have been different with respect to Göttling's *Almanach*, which, according to Martius, was "particularly influential in the education of apothecaries."[105] We therefore must take a closer look at this second chemical and pharmaceutical periodical that first appeared in 1780. It was published annually in the form of a small, bound book, its humble size matching its title *Almanach oder Taschenbuch für Scheidekünstler und Apotheker* (*Almanac or Pocketbook for Chemists and Apothecaries*).[106]

Göttling's *Almanach* addressed apothecaries primarily rather than "truly learned chemists [*würkliche gelehrte Scheidekünstler*]."[107] But as Göttling pointed out, it did "not contain recipes for the preparation of compound powders, spirits, pills, essences, tinctures, and ointments" – which, according to Göttling, belonged in a pharmacopoeia – as "its main goal was to make pharmacists [*Pharmaceutiker*] acquainted with proper chemical knowledge."[108] "Chemical knowledge" here meant, first of all, knowledge that was useful to "banish wrong treatment caused by ignorance, adulteration, the use of poisonous vessels, and many additional bad habits existing in apothecary's shops."[109] It also meant connoisseurship about a plethora of raw materials and chemical substances – the pharmacopoeias of the time enlisted between 2000 and 6000 different materials used as remedies—and the many ways of their identification and preparation. The practical pharmaceutical goal of learning more about the "genuineness of chemical products and the ways to avoid adulteration"[110] and the disinterested goal of acquiring learned knowledge about the realm of substances – a goal that accorded with the learned venture of experimental history (*historia experimentalis*) of the time[111] – were merely different sides of the same coin. In keeping with the empiricist attitude of Crell and the majority of German chemists, Göttling encouraged apothecaries to "pay attention in the future to incidents that often occur accidentally in apothecary's shops." He further emphasized that he did not expect all contributions to his journal to be "extended treatises," and that he was willing to publish "each single little remark."[112] Accordingly, he divided his journal into two main parts, the first of which contained "little remarks from chemistry" and the second one "extended treatises."

From the very beginning of its publication the *Almanach* seems to have been a great success. In a review of the *Almanach* of 1781, Crell observed that there was "strong demand" for the new journal.[113] He further remarked that this journal "promises to have the most pleasant consequences for our science [of chemistry], as it arouses and nourishes the germinating curiosity in such a pleasant manner."[114] Beginning in 1783, Göttling obtained an increasing number of contributions, especially by apothecaries who were not known as chemists. He included in his journal many letters and contributions by apothecaries that compiled a variety of different observations under headings such as "pharmaceutical and chemical remarks" and "mixed observations." In the eighth volume of the *Almanach*, published in 1787, Göttling proudly announced that readership of his journal was growing rapidly; he also reminded his readers that "during daily pharmaceutical-chemical work the attentive worker could always discover phenomena, which, even though they were economically useless, very often served to illuminate the darkest hypotheses by delivering facts; therefore I repeat that I am very grateful for the announcement of each observation."[115] Three years later, Crell wrote in another review that "this valuable publication continues to circulate a large amount of important and useful chemical knowledge rapidly and even to reach those [people] whose reading otherwise is very restricted, and often must be restricted."[116]

Among the well-known apothecary-chemists who contributed to Göttling's journal, J. C. Wiegleb and Johann Bartholomäus Trommsdorff (1770–1837) were the most active, apart from Göttling himself. But Göttling actually received contributions from many apothecaries, most of whom never became well-known chemists, such as Bindheim (Berlin), Böhme (Berlin), Borchers (Bückburg), Heyer (Braunschweig), C. A. Hoffmann (Weimar), Hoffmann (Ostfriesland), Remler (Erfurt), Liphardt (Finsterwalde), Merck (Darmstadt), Merkel (Nuremberg), Sprenger (Hanover), and Vogt (Erfurt). His request to report even minute observations of chemical phenomena obviously fell on willing ears, for in 1789 he received such an overwhelming number of contributions from amateurish apothecaries that he felt compelled to put on the brakes:

> In the diverse prefaces to this book I stated that every single observation submitted to me would be welcome, and as there is ample opportunity in *officines* to notice things by accident during the work that necessarily must be done, I have asked pharmacists to send me notices for publication and for the instruction of their colleagues. Several readers were inclined to do so from time to time, which I acknowledge with gratitude. I am also very pleased that through the publication of such notices I had the opportunity to encourage amateurs of chemistry (*Liebhaber der Scheidekunst*), who previously did not dare to present themselves in public. However during the ten years of editorship of this book, numerous letters and even entire papers have accumulated that are not of the least interest.

Almanach
oder
Taschen-Buch
für
Scheidekünstler und Apotheker
auf das Jahr 1784.

Fünftes Jahr.

Weimar
in der Hoffmannischen Buchhandlung.

Figure 6. Title-page of the *Almanach*

> Since such correspondence is not only time-consuming and troublesome, but also costs money, I am forced not to tolerate such letters any longer. I am willing to do justice to everybody: and because I also know very well that the first step is the hardest, I will not rest before I am able to extract the essence of even the most confused paper.[117]

Göttling closed his note with an example of the kind of letters he had in mind, in which the correspondent complained in very bad German and confused style about dangerous, smoky laboratories and praised the way smoke was drawn away in the laboratories of foundries (*Hütten Laboratorien*).

Göttling's *Almanach* was a forum for discussing the deficiencies of pharmaceutical apprenticeship and its reform, and it further addressed many issues of interest to apothecaries and chemists alike. An analysis of its content shows that the bulk of notes and papers concerned two activities: ways of unambiguously identifying substances, and improvements to techniques of preparing chemical remedies.[118] The unambiguous identification of chemical substances was part and parcel of the standardization of remedies and a prerequisite to avoid adulteration and frauds by merchants (*Materialisten*). Improvement to chemical manufacture meant economic gains and greater safety through preventing explosions and the use of poisonous vessels. But the technical activities linked with these goals also allowed shifts toward experimental histories of substances and their chemical analysis. Connoisseurship about thousands of substances, know-how of chemical preparation, analytical skills, accuracy in weighing, and conceptual chemical knowledge about composition were useful tools to the pharmaceutical art and at the same time part of learned chemical knowledge. But Göttling's journal was also concerned with other forms of learned chemical knowledge. For example, in the late 1780s it became a forum for chemists' discussion of the phlogistic and antiphlogistic theories and the new Lavoisierian chemical nomenclature. Since the *Almanach* also published reports about metallurgical and technological chemistry, it provided insight into many different areas of chemical practice and theory. In so doing, it contributed to a broader education of apothecaries in chemistry beyond what was immediately useful for pharmacy.

I argued above that Crell's journal played a significant role not only in the formation of the German chemical community, as highlighted by Hufbauer, but also in the emergence of apothecary-chemists. Even more so did Göttling's *Almanach*, and from 1794 Trommsdorff's *Journal der Pharmacie*, both of which fostered chemistry and addressed apothecaries more explicitly than Crell's journal. My analysis of the style of Göttling's *Almanach* – especially the fact that he encouraged apothecaries to report each little observation made during their daily work – shows that he encouraged apothecaries to read the journal and to contribute to it actively. Addressing both apprenticed apothecaries and learned chemists, Göttling integrated a considerable share of apothecaries into the community of German chemists. Furthermore all three journals amalgamated useful information with learned chemical knowledge and thus targeted the experience and interests of a heterogeneous group of readers. As contributors to Göttling's *Almanach,* Crell's *Chemische Annalen*, and

Trommsdorff's *Journal der Pharmacie*, apothecaries participated in the "regular communication" of the community of chemists.[119] I would like to extend Hufbauer's argument to include not only communication and justification, but also apothecaries' investigative practice in the laboratory. The possibility of publishing papers may have provided an incentive for more careful observation in the pharmaceutical laboratory and more systematic and extended chemical investigation. It may have reinforced a trend, which hinged on the correspondence between the material culture of the pharmaceutical and the chemical laboratories, to shift from the pharmaceutically useful to chemical analysis and the experimental history of substances.[120] Thus chemical journals may have fed back into the actual practice of apothecaries and contributed to the emergence of the persona of the apothecary-chemist on many different levels.

THE EMERGENCE OF APOTHECARY-CHEMISTS

Thus far we have discussed the system of pharmaceutical apprenticeship in eighteenth-century Germany, the acquisition of chemical knowledge by travels, and the role played by chemical journals both as a forum for communication between the heterogeneous groups of chemical investigators and as an institution that reinforced the possibility to shift from pharmaceutical manufacture to chemical analysis and the experimental histories of substances. Now we will discuss more closely the characteristics of the hybrid persona of the apothecary-chemist along with additional social and political conditions for its emergence in eighteenth-century Germany. Several features of this persona have already been highlighted above. German apothecaries' apprenticeship fit into the eighteenth-century system of apprenticeship in the arts and crafts. But pharmaceutical apprentices also knew some Latin, read books, performed series of chemical operations, and acquired skills to use the balance and to identify thousands of materials. Traveling as a journeyman was part of pharmaceutical learning, which also offered the possibility of extension to Enlightenment forms of travel. As a heuristic device, I proposed that four different groups of eighteenth-century German apothecaries be distinguished. This paper focuses on group four, that is, apothecaries who performed chemical operations, read chemical books and journals, contributed to chemical periodicals, and received acknowledgement as chemists in the community of German chemists or were even among their leading elite. Only in regard to these apothecaries, who defy a classificatory divide into either apothecaries or chemists, do I use the designation apothecary-chemists. In the following section I describe the professional career of two famous apothecary-chemists, Andreas Sigismund Marggraf and Martin Heinrich Klaproth, to further illuminate the persona of the apothecary-chemist and the collective conditions that contributed to its form of life.

A. S. Marggraf (1709, Berlin – 1782, Berlin) was the son of a grocer (*Materialist*) who had become member of the Berlin guild of grocers (*Materialistengilde*) in 1707, had received a privilege to buy the *Ratsapotheke* in 1720, and subsequently became one of the most renowned apothecaries of Berlin.[121] The first place of his apprenticeship was his father's apothecary's shop, followed in 1726 by the Royal *Hofapotheke*, administered at the time by Caspar Neumann. Marggraf's apprenticeship benefited from training by a knowledgeable apothecary-chemist like Neumann as well as from the newly established *Collegium medico-chirurgicum*, where he may have attended lectures on chemistry and practical chemical instruction by Pott and Neumann. After completing his apprenticeship in 1731, he served a year as journeyman in an apothecary's shop in Frankfurt-am-Main, and another year in J. J. Spielmann's *Hirsch-Apotheke* in Strasbourg. In 1733, he proceeded to Halle, where he took lectures at the university on medicine and on chemistry by Friedrich Hofmann (1660–1742) and Johann Juncker (1679–1759). A year later he went to the famous mining town of Freiberg to learn metallurgy and assaying with the renowned Mining Councillor Johann Friedrich Henckel (1678–1744). In 1735, he returned to Berlin by way of the Harz mining district, to become administrator of his father's apothecary's shop for 17 years.

Three years after his return to Berlin, the apprenticed apothecary A. S. Marggraf became an ordinary member of the Berlin Society of Sciences. In 1740, he published his first paper in the society's *Miscellanea Berolinensia*. During his tenure as administrator of his father's apothecary's shop (1740–52) he published a total of 15 reports on diverse chemical experiments – including experiments on phosphorus and its compounds (1740 and 1743), the precipitates formed from solutions of metals (1745), the extraction of zinc from calamine (1746), the dissolution of silver and mercury in vegetable acids (1746), the analysis of a salt obtained from urine (1746), dissolution of zinc in vegetable acids (1747), the extraction of sugar from beets and other plants (1747), the preparation of pure silver (1749), on luminescent stones (1749 and 1750), and on the oils extracted from insects (1749).[122] All of these experiments, including his most famous ones on the extraction of sugar from beets in 1747, were performed in the pharmaceutical laboratory of his father's shop. All of them were classical chemical analyses or experiments that explored modes of chemical preparation and chemical properties of substances.

By coincidence, shortly after his father had sold his shop, Marggraf was offered a salaried position at the Berlin Academy of Sciences. In 1752, the Academy began to implement an old plan dating back to the 1720s and the reign of Frederick Wilhelm I, by constructing a building that housed a laboratory and a residence for the Academy's chemist. The new building was on Dorotheenstrasse 10, opposite the observatory of the Academy.[123] In 1754, King Frederick II appointed Marggraf director of this laboratory, and shortly thereafter Marggraf moved into the building to continue his experiments at this new, academic location. In 1760 the Prussian king further decided that Marggraf, who had never completed an education at a university, was to become the director of the Physical Class of the Academy. By then he was already a famous chemist, not least through the public lectures he gave from the 1740s until the 1770s as well as his chemical achievements. To the latter belonged the famous extraction of sugar from beets in 1747; the identification of the syrupy phosphoric acid as a peculiar acid in 1743; the improved preparation of phosphorus in the same year; and the identification of alumina as a peculiar alkaline earth in 1754. In a memoriam published in the *Chemische Annalen* in 1788, Crell celebrated Marggraf as the second greatest German chemist (after Stahl) and the "renewer" of the present "European chemistry."[124] In particular he praised the clarity of his experiments and his abstention from premature hypotheses and any "system." He further emphasized his high reputation in Britain and France, manifested by his election as one of the six foreign associates of the Paris Royal Academy of Sciences.[125]

A pharmaceutical apprenticeship in Berlin – where the Medical-Surgical College, the two laboratories of the Royal *Hofapotheke*, and the Enlightenment system of public teaching of science provided extraordinarily good opportunities for the learning of chemistry – the openness of the Berlin Society of Sciences to artisans, and the strong support of chemical practitioners by the Prussian state were the most important social prerequisites that contributed to Marggraf's career as a chemist. In the 1770s, Berlin hosted a second apothecary who would become an even more famous chemist than Marggraf: Martin Heinrich Klaproth (1743, Wernigerode – 1817, Berlin). Klaproth, the son of a tailor, was educated at the Latin school of Wernigerode, which he left in 1758 after four years without graduating.[126] From 1759 until 1764 he was a pharmaceutical apprentice in the *Ratsapotheke* of Quedlinburg (see above), where he learned chemical techniques but apparently nothing that went beyond pharmaceutical chemistry. Nevertheless, Klaproth spent the first two years of his service as a journeyman in this apothecary's shop, before proceeding to the *Hofapotheke* of Hanover. This *Hofapotheke* had an outstanding reputation, as it was owned by August Hermann Brande, member of a well-known family of apothecaries and equipped with a good laboratory and a library. Apart from Klaproth, another famous apothecary-chemist was trained there at the time, namely Johann Friedrich Westrumb (1751–1819), who left some descriptions of the place. According to Westrumb, Klaproth made extensive use of the library and its newest books on chemistry, such as those by J. R. Spielmann (*Institutiones chemiae*, 1763) and F. Cartheuser (*Elementa chymiae*, 1736). He further used the laboratory to perform chemical operations, especially experiments that were not of immediate pharmaceutical use.[127] In 1768, Klaproth traveled to Berlin, the Prussian center for advanced chemical–pharmaceutical learning. In Berlin he found a position in one of the best of its 22 apothecary's shops, the large and modern *Apotheke zum Engel* at Mohrenstraße 5.[128] He then completed his total of seven years of service as a journeyman in 1770 in the *Ratsapotheke* of Danzig, and returned to Berlin in 1771 to accept a position in the *Apotheke zum weißen Schwan*, owned by the apothecary-chemist Valentin Rose (1736–71). When Rose died shortly afterwards at the age of 35, Klaproth became administrator of his apothecary's shop after he had completed the required examination by the *Ober-collegium-medicum*.[129]

In 1780, through his marriage to a niece of Marggraf, Klaproth came into sufficient funds to buy his own apothecary's shop. In the two decades that followed, his *Apotheke zum Bären* on Spandauer Straße prospered both economically and scientifically. The only publication by Klaproth before 1780 was included in a volume published in 1776 by the Berlin physician Marcus Elieser Bloch (one of the founders of the *Gesellschaft naturforschender Freunde zu Berlin* in 1773), where a footnote by Bloch mentioned that its author was "a certain *Herr* Klaproth" who administered Rose's apothecary's shop. However, from 1782 until 1800, when Klaproth sold his shop, he published a total of 84 papers based on experiments performed in the laboratory of his apothecary's shop.[130] As Dann pointed out, after the death of the famous apothecary and chemist Carl Wilhelm Scheele (1742–86), Klaproth's apothecary's shop and laboratory became the most productive artisanal site of scientific chemical investigation in all of Europe. In his pharmaceutical laboratory Klaproth discovered the elements uranium and zirconium (in the form of oxides) in 1789, rediscovered titanium (in the form of titanium dioxide) in 1795 and chromium in 1798, confirmed the existence of strontia in 1793, and established the identity of tellurium and chromium in 1798. At the same time, Klaproth was a very successful entrepreneur who transformed his apothecary's shop into the second largest in Berlin, receiving three times its original price when he sold it in 1800.

In the 1780s, Klaproth's reputation as a knowledgeable chemist increased rapidly. His public lectures on chemistry from 1782 on became the newest fashion among Berlin's intellectual elite (see the introduction). In 1782, he took a position as private lecturer at the *Collegium medico-chirurgicum*, followed by a salaried teaching position in 1784 at the Berlin Mining School (*Bergschule*, founded in 1770), and another teaching position in 1787 at the artillery school of General G. F. von Tempelhoff, which became the Royal Artillery Academy in 1791. In 1788, Klaproth became an unsalaried member of the Berlin Academy of Sciences. In addition he assumed positions as a Prussian medical official, councillor, and commissioner; as commissioner he investigated the Royal Porcelain Works with A. v. Humboldt in 1792, and reviewed Franz Carl Achard's experiments for the extraction of sugar from beets, some of which were performed in Klaproth's laboratory in the *Apotheke zum Bären*. In 1800, after Franz Carl Achard (1753–1821) had retired from his salaried position at the Academy of Sciences to establish a sugar beet factory, Klaproth became the director of the laboratory of the Berlin Academy of Sciences.[131] After Marggraf and Achard, he was the third salaried chemist in a row at the Academy of Sciences with no university degree. Thereupon he sold his apothecary's shop in the same year to take up residence in the building on Dorotheenstraße 10 that hosted the Academy's laboratory. Since in the previous year Achard had used the Academy's laboratory as a technological station to extract sugar from beets on a large scale, however, the laboratory was rotting due to the inevitable results of the heavy production of beet sugar.[132] Klaproth's complaints resulted in the construction of a new laboratory on Dorotheenstrasse that was finished in December 1802.[133] When Klaproth moved into the new building, the equipment of his old pharmaceutical laboratory moved with him—that is, as in the case of Marggraf's move there was a direct transfer of instruments, vessels, and materials from the pharmaceutical to the academic laboratory.[134] In 1810 Klaproth reached the pinnacle of his scientific career by becoming the first ordinary professor of chemistry on the philosophy faculty of the newly founded University of Berlin. By this time Klaproth's reputation as an eminent chemist was manifest not only in his audience, which also included distinguished German philosophers such as Arthur Schopenhauer and Friedrich D. E. Schleiermacher, but also by his membership in the Royal Society of London (1795) and the Paris Academy of Sciences (1804).

Marggraf and Klaproth were both apprenticed apothecaries who had never earned a university degree, and for a long period administered or owned an apothecary's shop and manufactured and sold remedies and other commodities; other renowned apothecary-chemists of the time such as J. C. Wiegleb and J. F. Westrumb never left their apothecary's shop. The two apothecary-chemists were also salaried members of the Berlin Academy of Sciences and directors of its chemical laboratory – and Marggraf even the director of its Physical Class; members of other scientific societies in Germany and abroad; authors of numerous texts on chemical science covering a broad range of chemical subjects; teachers of chemistry, and, in the case of Klaproth, a professor of chemistry at the Friedrich Wilhelm University of Berlin. Their careers as learned chemists did not begin after they had finished their pharmaceutical business but rather developed alongside, and even in conjunction with that business. There is no evidence whatsoever that they performed chemical experiments in their laboratories at the expense of pharmaceutical manufacture or otherwise neglected their pharmaceutical business. Rather, they were both enterprising apothecaries and scientific chemists, or apothecary-chemists.

It goes without saying that the career of outstanding apothecary-chemists such as Marggraf and Klaproth owed much to specific local circumstances and individual talent. But these individuals could only succeed in a culture and system of institutions that provided the resources for their success. Marggraf and Klaproth were not unique geniuses who had to bridge a huge gap between a rigid "realm of recipes" and routine and a realm of progressive chemical science.[135] Such a gap did not exist in the eighteenth century. Marggraf and Klaproth rather stood on top of an iceberg, whose invisible parts comprised the many German apothecaries who performed chemical investigations alongside pharmaceutical manufacture. Moreover, they were hardly alone in the visible area of that iceberg, as there were other renowned apothecary-chemists in eighteenth-century Germany such as Neumann, Wiegleb, Westrumb, F. A. C. Gren (1760–98), V. Rose Sr. (1736–71), V. Rose Jr. (1762–1807), Trommsdorff, and Hermbstädt. The outstanding achievements of one Marggraf and one Klaproth were embedded in a collective pharmaceutical practice that had implemented chemical instruments, techniques, reagents, and chemical preparations in the course of the seventeenth century, and that objectively allowed a shift from the production of chemical remedies to chemical analysis and the experimental history of substances. Their achievements further hinged on a public culture and scientific institutions that were open to certain groups of practitioners such as apothecaries and cameralists, as well as on policies by the mercantilist German governments that fostered the scientific careers of skilled practitioners.[136] The prospects of publishing articles in learned journals, of election as a member of scientific societies and academies, of a professorship at the new professional schools or even at a university, or of giving lectures in the system of broader public teaching – all of these provided social incentives for scientific chemical investigation. Moreover, in eighteenth-century Germany, state intervention, the accessibility of scientific institutions by practitioners, and cultural interest in the experiential and useful sciences met with a pharmaceutical art that was internally innovative and whose material culture strongly overlapped with the material culture of the chemistry performed at academic sites. Marggraf's and Klaproth's pharmaceutical laboratories were both sites of commercial production and scientific investigation; likewise, the Berlin Academy's laboratory was not only a place for chemical experiments but also for large-scale technological inquiry into the extraction of sugar from beets. The emergence of the hybrid persona of apothecary-chemist in eighteenth-century Germany was significantly conditioned by the confluence of these diverse social, cultural and material conditions.

CONCLUSION

The role played by skilled practitioners, as compared to philosophers and savants educated at universities, in the development of the experimental sciences from the seventeenth until the nineteenth century has been an issue debated among historians of science for decades. Historians of physics have argued that ancient conventions of separating "authorship from experimental hands" continued well into the eighteenth century and beyond.[137] This view – highlighted by Steven Shapin's distinction between gentlemanly philosophers and "invisible technicians" in seventeenth-century England[138] – requires modifications, however, when the history of chemistry is involved. In late seventeenth- and eighteenth-century chemistry, apothecaries played a significant role, not merely as skilled technicians but also as authors of chemical texts, members of scientific societies and academies, public lecturers and teachers at professional schools and universities. All of these facts would speak for their classification as learned men (later "scientists"). Yet at the same time, eighteenth-century Germany apothecaries, as rule, were trained in the artisanal system of apprenticeship. And even apothecaries who became famous as chemists carried out chemical experiments alongside pharmaceutical manufacture and administration of the *officine*. Using criteria such as education and training, occupations, authorship, membership in scientific institutions, and social acknowledgement we have to conclude that there was a considerable group of apothecaries in eighteenth-century Germany that fits into both the group of artisans and that of learned men (or scientists).[139]

When leading American historians of science back in the 1950s and 60s criticized earlier attempts to put the Scientific Revolution into the context of developments in technology, they especially attacked the "craftsman-and-scholar thesis,"[140] which Rupert Hall summarized as follows:

> In any case, I hesitate to conclude that the behavior of an empirical scientist – that is, I take it, one who observes and experiments, both to discover new information and to confirm his statements and ideas – is derivable by virtually direct imitation from the trial-and-error, haphazard, and fortuitous progress of the crafts. This seems to me to be the defect of the view that sees the new scientist of the seventeenth century as a sort of hybrid between the older natural philosopher and the craftsman.[141]

Hall argued that the Scientific Revolution – which he regarded primarily as an innovation of the sciences that were established at universities such as astronomy, natural philosophy, and anatomy – was primarily the result of "internal strife" between innovators and conservatives within the academic world.[142] Recent historical research, according to Hall, had shown that technology and skilled practitioners had only limited significance for the Scientific Revolution. With respect to the experimental sciences, he demarcated "empirical scientists" from skilled practitioners who were not educated at universities by depicting the former as disinterested observers of nature and lumping together the latter as unlettered, routine, and machine-like craftsmen (see the quotation above). In the 1970s, Thomas Kuhn elaborated Hall's argument by introducing the distinction between "classical sciences" and "Baconian sciences." Like Hall, Kuhn regarded the transformations of the university-based classical sciences during the Scientific Revolution mainly as the result of learned men's "new ways of looking at old phenomena."[143] But he also argued that, largely independent of the transformations in the classical sciences, such as astronomy, geometrical optics, and statics, a second branch of sciences developed outside the universities and other academic institutions, the experimental "Baconian sciences." He further asserted that well into the nineteenth century the development of the Baconian sciences was spurred by "amateurs," that is, men not educated at universities and not participating in the world of "learned discourse."

Yet both Kuhn and Hall made an interesting exception.[144] Hall conceded that "chemistry reveals a very different historical pattern, in which almost everything said of astronomy [as an example of academic sciences] is negated."[145] He further pointed out that "craftsmen had developed both qualitative and quantitative techniques of vital necessity to the growth of chemistry as an exact science."[146] Likewise, Kuhn stated that "except in chemistry, among pharmacists and doctors, actual practice was seldom combined with learned discourse."[147] With respect to the institutional setting of the Baconian sciences, Kuhn "excepted" chemistry as well;[148] as is well known, as early as the seventeenth century chemists were elected as members of the Royal Society and of the Paris Academy of Sciences, and received professorships at the Paris *Jardin du Roy* and the medical faculties of universities.[149] But why was chemistry an "exception," even within the Baconian sciences? Kuhn was silent about his criteria for this judgement, which implied, of course, that chemistry was irrelevant for his larger picture of the Baconian sciences, whose centerpiece was experimental physics ("experimental philosophy"). I argue that any attempt to evaluate in a historically adequate way the importance, social and epistemological, of chemistry vis à vis experimental physics in the eighteenth century must take into account that chemistry was the first experimental science that was more broadly accepted at early modern academic institutions. Early modern chemistry also established a sustained experimental practice and a specific site of experimentation, namely, laboratories.[150] Compared to the number of experimental philosophers who investigated electricity, magnetism, and heat, that is, fields of inquiry later constituting experimental physics, there were large communities of chemists in the eighteenth century, especially in France, Germany, and Sweden. Furthermore, compared to the number of experimental reports on experimental physics, there were legions of experimental reports dealing with chemistry at the time. All of this manifests clearly that eighteenth-century chemistry was by no means a marginal experimental science. On the contrary, there was quite significant social and intellectual support for the experimental science of chemistry, which was more involved in the mundane business of dirty handiwork and uncanny materials than any other early modern experimental discipline. Rethinking the importance of chemistry in the landscape of early modern learned inquiries into nature may considerably improve our understanding of the social roots of the modern sciences and their relationships with technology and society.

NOTES

1 Unless noted, all translations are my own. The anonymous letter entitled "Vom Herrn M. H. in Berlin," was published in the *Chemische Annalen für die Freunde der Naturlehre, Arzneygelahrtheit, Haushaltungskunst und Manufakturen*, pt. 1, 1784, 342. None of the subscribers to the *Chemische Annalen* who lived in Berlin at the time had the initials M. H., see Karl Hufbauer, *The Formation of the German Chemical Community (1720–1795)* (Berkeley: University of California Press, 1982), 293 and 278–80.

2 See Georg Edmund Dann, *Martin Heinrich Klaproth, 1743–1817* (Berlin: Akademie-Verlag, 1958), 65f.

3 Hufbauer, "German Chemical Community," 85.

4 Lorenz Crell, "Vorrede," *Chemisches Journal für die Freunde der Naturlehre, Arzneygelahrtheit, Haushaltungskunst und Manufacturen*, pt. 1, 1778, 9–20, on 9; translated in Hufbauer, "German Chemical Community," 70.

5 Crell, *Chemisches Journal*, pt. 1, 1778, 9–20, on 10. Hufbauer quoted additional claims by eighteenth-century German chemists to leadership among European chemists (Hufbauer, "German Chemical Community," 83). Although German chemists' self-assessment is rhetorical, it should be noted that the strong focus of historians of chemistry on French chemistry in the eighteenth century is somewhat unbalanced from a historical point of view. In the late eighteenth century, France and Germany, as well as Sweden, were regarded as countries where chemistry flourished. For example, in 1791 the translators of Crell's *Chemische Annalen* into English mused about the "cause of the great(er) progress of chemistry in France, Germany and Sweden"; see *Crell's Chemical Journal* 1, 1791, 106; *Chemische Annalen*, pt. 2, 1778, 188–91, on 188.

6 Hufbauer, "German Chemical Community," 54. Between 1720 and 1780 Hufbauer counted 42 renowned chemists. He estimated that this number represented a fifth or sixth of all Germans carrying out chemical investigation. For his criteria of selection of well-known chemists, see Hufbauer, "German Chemical Community," 50. On chemists and apothecaries in late eighteenth-century Germany, see also Bernard Gustin, *The Emergence of the German Chemical Profession 1790–1867* (PhD thesis: The University of Chicago, 1975); Ernst Homburg, "Two Factions, One Profession: The Chemical Profession in German Society 1780–1870," 39–76 in *The Making of the Chemist: The Social History of Chemistry in Europe 1789–1914*, David Knight and Helge Kragh, eds. (Cambridge: Cambridge University Press, 1998).

7 Hufbauer, "German Chemical Community," 54–55.

8 Several of these chemists would later obtain positions as mining and metallurgical officials. For additional arts and crafts in Germany that implemented chemical techniques and knowledge, see Hufbauer, "German Chemical Community," 57–61; Ursula Klein, "Technoscience avant la lettre," *Perspectives on Science* 13, 2005, 227–66.

9 In France, a professional school was established in 1777, the *Collège de Pharmacie de Paris*, and apothecaries were officially designated *pharmacien* from that time. See Edward Kremers and Georg Urdang, *History of Pharmacy: A Guide and a Survey* (Philadelphia: J. B. Lippincott, 1940); Jonathan Simon, *Chemistry, Pharmacy and Revolution in France, 1777–1809* (Aldershot: Ashgate, 2005); Sacha Tomic, *Les Pratiques et les Enjeux de l'Analyse Chimique des Végétaux: Étude d'une Culture Hybride (1790–1835)* (PhD diss., Université de Paris X–Nanterre, 2003).

10 See Hufbauer, "German Chemical Community," 86–88. Information about subscribers for the period 1784–89 is given in Crell's journal.

11 The most active contributor (68 papers) was the apothecary and chemist Johann Friedrich Westrumb (1751–1819), who first administered the *Hofapotheke* in Hanover and was then lessee of the *Ratsapotheke*

in Hameln until his death in 1819. Westrumb performed all the experiments he reported in his publications in his pharmaceutical laboratory. In addition he was mining commissioner, member of the Chamber of Commerce of Hannover, and a chemical entrepreneur who attempted to establish commerical bleaching with chlorine from 1789 to 1790. Hufbauer, "German Chemical Community," 205.

12 Relying strongly on the rhetoric of the Lavoisier group, Simon's "Chemistry," argues that the Chemical Revolution engendered a bifurcation of pharmacy and chemistry; Tomic, *Les Pratiques* questions this view.

13 Among the areas of production that intersected with chemistry, the most important were pharmacy and metallurgy, especially assaying. The terms "technology" (*Technologie*, which means systematic knowledge about technical devices and the arts and crafts, or industry) and "chemical technology" were used by German chemists from ca. 1770 onward.

14 It should be noted that my focus on the apothecary-chemists does not entail the assertion that these men were more important than other prominent eighteenth-century German chemists who were educated mainly at universities and predominantly professors teaching at universities, such as Georg Ernst Stahl (1659–1734), Friedrich Hoffmann (1660–1742), Johann Eller (1689–1760), Johann Heinrich Pott (1692–1777), Johann Friedrich Gmelin (1748–1804), and Christian Ehrenfried Weigel (1748–1831). Furthermore my focus in this paper on people and social institutions implies that I will not study in detail other interesting features of the interconnectedness of chemistry and pharmacy in eighteenth-century Germany, such as the overlapping material culture and the correspondence between techniques of producing chemical remedies and techniques of chemical analysis. It further implies that I will not examine in detail the questions of what different forms of knowledge existed in eighteenth-century chemistry and which parts of it played a role in the pharmaceutical handicraft, and which were irrelevant. These additional aspects will be studied in two forthcoming papers: Ursula Klein, "Apothecary's Shops, Laboratories and Manufacture in Eighteenth-Century Germany" 246–276 in *The Mindful Hand: Inquiry and Invention from the Late Renaissance to Early Industrialization*, Lissa Roberts, Simon Schaffer, and Peter Dear, eds. (Amsterdam: Royal Netherlands Academy of Arts and Sciences, forthcoming) and "Blending Technical Innovation and Learned Natural Knowledge: The Making of Ethers" in *Between the Marketplace and the Laboratory: Materials and Expertise* (1500–1800), Ursula Klein and Emma Spary, eds., forthcoming.

15 For the stock of remedies that existed in Germany before the Paracelsian movement, see Dietrich Arends, Erika Hickel, and Wolfgang Schneider, *Das Warenlager einer mittelalterlichen Apotheke (Ratsapotheke Lüneburg 1475)* (Braunschweig: Technische Hochschule, 1960); Astrid Müller-Grzenda, *Pflanzenwässer und gebrannter Wein als Arzeneimittel zu Beginn der Neuzeit: Herstellungsverfahren, Hersteller und Handel, Beschaffenheit und Bedeutung für die Materia Medica* (Stuttgart: Deutscher Apotheker Verlag, 1996); Erika Hickel, *Salze in den Apotheken des 16. Jahrhunderts* (Braunschweig: Technische Hochschule, 1965); R. J. Forbes, *Short History of the Art of Distillation, from the Beginnings up to the Death of Cellier Blumenthal* (Leiden: Brill, 1948); Hermann Schelenz, *Zur Geschichte der pharmazeutisch-chemischen Destilliergeräte* (Hildesheim: Georg Olms, 1964).

16 "Pharmacopoeias" (or "dispensatories") were apothecary books with governmental authority; in Germany, their authors were physicians until the late eighteenth century, when apothecaries also became authors of pharmacopoeias. *Arzeneitaxen* were lists of fixed prices for medicines set by the government.

17 The Prussian medical edict of 1693 is reprinted in Manfred Stürzbecher, *Beiträge zur Berliner Medizingeschichte: Quellen und Studien zur Geschichte des Gesundheitswesens vom 17. bis zum 19. Jahrhundert* (Berlin: Walter de Gruyter, 1966), 43–64, see 49; on the term "laboratory" see 51 and 54. The term "laborant" meant distillers, many of whom came from Thüringen. The 1685 Brandenburg medical edict (also reprinted in Stürzbecher, *Beiträge*, 27–34) explicitly excluded alchemists from the preparation of chemical remedies, except those for which they had obtained privileges (see 31). For a similar order in the 1721 medical edict of the Duchy Brunswick-Wolfenbüttel see Gabriele Beisswanger, *Arzeneimittelversorgung im 18. Jahrhundert: Die Stadt Braunschweig und die ländlichen Distrikte im Herzogtum Braunschweig-Wolfenbüttel* (Braunschweig: Braunschweiger Veröffentlichungen zur Geschichte der Pharmazie und der Naturwissenschaften, 1996).

18 See Erika Hickel, "Der Apothekerberuf als Keimzelle naturwissenschaftlicher Berufe in Deutschland," *Medizinhistorisches Journal* 13, 1978: 259–76; Erika Hickel, *Apotheken, Arzeneimittel und Naturwissenschaften in Braunschweig*, 1677–1977 (Braunschweig: Hagenmarkt-Apotheke, 1977); Erika Hickel, *Arzeneimittel-Standardisierung im 19. Jahrhundert in den Pharmakopöen Deutschlands, Frankreichs, Großbritanniens und der Vereinigten Staaten von Amerika* (Stuttgart: Wissenschaftliche Verlagsgesellschaft, 1973); Wolfgang Schneider,

Geschichte der pharmazeutischen Chemie (Weinheim: Verlag Chemie GmbH, 1972) and *Lexikon zur Arzeneimittelgeschichte: Sachwörterbuch zur Geschichte der pharmazeutischen Botanik, Chemie, Mineralogie, Pharmakologie, Zoologie*, 7 vols. (Frankfurt a. M.: Govi-Verlag, 1968–74); Mechthild Krüger, *Zur Geschichte der Elixiere, Essenzen und Tinkturen* (Braunschweig: Technische Hochschule, 1968); Beisswanger, *Arzeneimittelversorgung*.

19 For general outlines of the pharmaceutical apprenticeship in eighteenth-century Germany see Kremers and Urdang, *History*, 120–22; Alfred Adlung and Georg Urdang, *Grundriß der Geschichte der Deutschen Pharmazie* (Berlin: Julius Springer, 1935), 133–34; Berthold Beyerlein, *Die Entwicklung der Pharmazie zur Hochschuldisziplin (1750–1875), ein Beitrag zur Universitäts- und Sozialgeschichte* (Stuttgart: Wissenschaftliche Verlagsgesellschaft, 1991).

20 The *officine* was the public room of an apothecary's shop, where the apothecary dispensed Galenic remedies and sold all kinds of remedies; see also Figure 3, which shows the many rooms belonging to an early modern apothecary's shop.

21 Stürzbecher, *Beiträge*, p. 54.

22 Ibid.

23 Ibid., 52. *Purgantia* were Galenic medicines for "purification."

24 I will come back to this point below.

25 The school was founded in 1724 primarily for the education and training of surgeons, especially military surgeons. See Herbert Lehmann, *Das Collegium medico-chirurgicum in Berlin als Lehrstätte der Botanik und der Pharmazie* (Berlin: Triltsch & Huther, 1936).

26 The relevant passage of the medical edict is published in ibid., 14.

27 At the same time, the *collegium medicum* in Berlin was reorganized as *Ober-collegium-medicum*, which was at the top of the hierarchy in the medical system and controlled the *collegia medica* in the provinces.

28 Lehmann, *Collegium*, 10, 18. The buildings of the Berlin Society of Sciences belonged to a complex of buildings, which also included the Royal Marstall, located between *Unter den Linden* and *Dorotheenstraße*.

29 See Caspar Neumann, *Chymia medica dogmatico-experimentalis* (Züllichau, 1756).

30 See Alexander von Lyncker, "Die Matrikel des preußischen Collegium medico-chirurgicum in Berlin 1730 bis 1768," *Archiv für Sippenforschung* 11 (Heft 5), 1934, 129–57.

31 See Lehmann, *Collegium*, 16; Hickel, "Apothekerberuf"; Adlung and Urdang, *Grundriß*, 134.

32 Wiegleb's autobiographical notes were integrated into a necrology by his friend, the physician Dr. Stoeller; see Stoeller, "Nekrolog Johann Christian Wiegleb," *Allgemeines Journal der Chemie* 4, 1800, 684–720. On Wiegleb see also Fritz Krafft, "Johann Christian Wiegleb und seine Rolle bei der Verwissenschaftlichung der Chemie," 151–95 in *Apotheker und Universität: die Vorträge der Pharmaziehistorichen Biennale in Leipzig vom 12. bis 14. Mai 2000*, Christoph Friedrich and Wolf-Dieter Müller-Jahncke, eds. (Stuttgart: Wissenschaftliche Verlagsgesellschaft, 2002).

33 Stoeller, "Nekrolog," 689.

34 Ibid.

35 In addition he bought Gottfried Rothe's *Gründliche Einleitung zur Chemie* (1717); Johann Kunckel's posthumously published *Laboratorium Chymicum* (1716); Christoph Heinrich Keil's *Compendiöses, Doch vollkommenes Medicinisch-Chymisches Handbüchlein* (1734); and the metallurgical and chemical writings of Georg Ernst Stahl (ibid., 690).

36 In 1782, Wiegleb also published a new, abridged translation of Herman Boerhaave's *Elementa chemiae* (1732), which concentrated on the practical part of the book.

37 Stoeller, "Nekrolog," 691.

38 Kremers and Urdang quote Goethe to highlight the high social esteem of German apothecaries around 1800. In 1822 Goethe stated that "in Germany the apothecary enjoys a highly esteemed position within society … The German apothecaries cultivate science. They are aware of its importance and endeavor to utilize it in practical pharmacy"; Kremers and Urdang, *History*, 123.

39 Stoeller, "Nekrolog," 693.

40 Ibid., 696.

41 Hufbauer, *German Chemical Community*, 191.

42 Stoeller, "Nekrolog," 696.

43 For boarding schools see also below.

44 "Auch ein kleiner Beytrag über den Zustand der Pharmacie in Deutschland," *Almanach oder Taschenbuch für Scheidekünstler und Apotheker* 1793, 49–72; Göttling's comment is 72–75.

45 Anonymous, "Kleiner Beytrag," 57–63.

46 See Dann, *Klaproth*, 19–20; Grünhagen, "Einrichtung Apotheken"; Hermann Lorenz, *Die Ratsapotheke zu Quedlinburg* (transcript of the *Quedlinburger Kreisblatt*, Nr. 178–81, 1928, courtesy of the *Universitätsbibliothek Freie Universität Berlin*). In addition to the rooms mentioned above, there were large rooms and smaller rooms on the first floor for guests to town and for the journeyman, and private rooms for the apothecary on the ground floor. The apprentices had to sleep in a large closet next to the laboratory.

47 For the chemical remedies that were newly introduced in Germany in the period 1670–1750, see Schneider, *Lexikon*, 3: 75–127 and *Geschichte*, 148–52.

48 A document from 1754 mentions, in particular, six furnaces for distilling, two furnaces for the manufacture of spirits, a chimney with a mantle, and instruments for cupellation. See Konrad Grünhagen, *Über den Bau und die Einrichtung von Apotheken in alter und neuer Zeit* (Würzburg: Konrad Triltsch, 1939), 79.

49 In the 1780s the *Ratsapotheke* of Quedlinburg was run by an apothecary named Schacht who was also a subscriber to Crell's *Chemische Annalen* (see Hufbauer, "*German Chemical Community*," 287).

50 Quoted in Dann, *Klaproth*, 15.

51 Dann, *Klaproth*, 15 and 142. His grandfather, father, and uncle were physicians; his father wrote an alchemical text on tincture of gold.

52 Johann Christian Lüderitz Liphardt, "Bemerkungen, Wünsche und Vorschläge für sämtliche Herren der Apothekerkunst; als ein Nachtrag zur moralischen Disziplin des Herrn Bindheim," *Almanach* 1784, 70–98, on 73. Liphardt's article was a response to an article by the Berlin apothecary Johann Jacob Bindheim, published in the previous issue of the *Almanach* of 1783. In this article Bindheim had argued for the traditional moral virtues of an apprentice, which also had been required in the Prussian medical edict (see above). Bindheim demanded above all that the pharmaceutical apprentice be obedient, hardworking, and neat; apart from some knowledge of Latin, good orthography, and the readiness to read books, he did not stress the need for a more learned education, as Liphardt did. See Johann Jacob Bindheim, "Sendschreiben, über die moralische Disciplin des Apothekers," *Almanach* 1783, 81–93. On Bindheim and Liphardt see also Wolfgang-Hagen Hein and Holm-Dietmar Schwarz, *Deutsche Apotheker-Biographie*, 4 vols. (Stuttgart: Wissenschaftliche Verlagsgesellschaft, 1975–97).

53 Liphardt, "Bemerkungen," 73.

54 Ibid., 84.

55 Ibid., 89–90.

56 Ibid., 86.

57 Ibid., 91.

58 On the different eighteenth-century social groups occupied with the production of remedies – mostly small, specific groups of them such as distilled aromatic oils, alcoholic spirits, and syrups – see also Beisswanger, *Arzeneimittelversorgung*; Ulla Meinecke, *Apothekenbindung und Freiverkäuflichkeit von Arzeneimitteln: Darstellung der historischen Entwicklung bis zur Kaiserlichen Verordnung von 1901 unter besonderer Berücksichtigung des Kurfürstentums Brandenburg und des Königreichs Preußen* (PhD diss.: University of Marburg, 1971).

59 This also becomes manifest in the textbooks published by apothecary-chemists, which covered the entire range of practical or "applied chemistry" and chemical theory.

60 Martius was a subscriber to Crell's *Chemische Annalen* (see Hufbauer, *German Chemical Community*, 283), but he was a marginal figure in the German chemical community (ibid., 107); apart from his autobiography, he published four smaller works, three of which dealt with pharmaceutical botany and one with mineralogical observations; Hein and Schwarz, *Deutsche Apotheker Biographie*, 2: 409–10.

61 Ernst Wilhelm Martius, *Erinnerungen aus meinem neunzigjährigen Leben* (Leipzig: Leopold Voss, 1847), 17–19.

62 Göttling published an extract of the letter as "Auszug eines Schreibens über pharmaceutische Lehrmethode," *Almanach* 1787, 62–77.

63 See Karl Gottfried Hagen, *Lehrbuch der Apothekerkunst* (Königsberg, 1786), 5–6.

64 Hagen, *Lehrbuch*, 49.

65 Ibid., 51–52.

[66] "Lehrmethode," 67–71.

[67] Ibid., 73.

[68] See Bindheim, "Sendschreiben"; Liphardt, "Bemerkungen"; "Lehrmethode"; "Ueber den Zustand der Pharmacie in Wien. An meine Mitbrüder in Niedersachsen und den Reichslanden," *Almanach* 1792, 49–108; in reply to the former: "Auch ein kleiner Beytrag über den Zustand der Pharmacie in Deutschland," *Almanach* 1793, 49–74; Johann Bartholomäus Trommsdorff, "Der vollkommene Apotheker," *Almanach* 1793, 75–96 and "Methode, junge Leute zu brauchbaren Apothekern zu erziehen," *Journal der Pharmacie für Ärzte und Apotheker* 1, pt. 1, 1793, 29–39; "Ueber Hrn. Apotheker Merkels in Nürnberg Vorschläge, sich künftig brauchbarere Apotheker-Subjekte zu verschaffen. Von einem der Pharmacie Beflissenen," *Almanach* 1793, 119–32; J. R. Spielmann, "Ueber pharmaceutische Schulanstalten," *Journal der Pharmacie für Aerzte und Apotheker* 1, pt. 1, 1793, 48–58.

[69] "Apotheker-Subjekte," 112 and 122; "Kleiner Beytrag."

[70] Trommsdorff, "Apotheker," 80–81.

[71] Trommsdorff, "Plan und Zweck dieser Zeitschrift," *Journal der Pharmacie für Aerzte und Apotheker* 1, pt. 1, 1793, I–XII, II.

[72] Trommsdorff, "Methode."

[73] The boarding school, which addressed not only prospective apothecaries, and its curriculum were announced in Trommsdorff's journal; see Johann Bartholomäus Trommsdorff, "Nachricht von einer chemisch-physikalischen, und pharmaceutischen Pensionsanstalt, für Jünglinge," *Journal der Pharmacie für Aerzte und Apotheker* 2 (1795), part 1 (1794). As Hufbauer pointed out, this successful school was open until 1828, and taught more than 300 students, of which at least 13 subsequently established chemical factories. In 1823, the Prussian government announced that attending this school was equivalent to university attendance; see Hufbauer, *German Chemical Community*, 218ff.

[74] Sigismund Friedrich Hermbstädt, "Nachricht von einer chemischen Pensionsanstalt für Jünglinge, die sich zu praktischen Chemikern bilden wollen," *Chemische Annalen*, pt. 1, 1790, 94–96, on 94. On chemical and pharmaceutical boarding schools in eighteenth-century Germany see also Dieter Pohl, *Zur Geschichte der pharmazeutischen Privatinstitute in Deutschland von 1779–1873* (PhD diss., University of Marburg, 1972); Gustin, "German Chemical Profession," 60–77.

[75] Hermbstädt, "Nachricht," 96.

[76] See, for example, S. R. Epstein, "Craft Guilds, Apprenticeship, and Technological Change in Preindustrial Europe," *Journal of Economic History* 58, 1998, 684–713; Reinhold Reith, "Technische Innovationen im Handwerk der Frühen Neuzeit," 21–60 in *Stadt und Handwerk in Mittelalter und Früher Neuzeit*, Karl Heinrich Kaufbold and Wilfried Reininghaus, eds. (Köln: Böhlau, 2000).

[77] See Stoeller, "Nekrolog," 698.

[78] Martius, *Erinnerungen*, 38.

[79] Martius remarked that journeymen and clerks of apothecaries were treated "like commodities" at the time, "for when they needed a position they addressed a grocery (*Materialhandlung*), and the master apothecary (*Principal*) did the same when he wanted to fill the position of a clerk. The apothecary who bought most from the *Materialist* got the best clerk." (Martius, *Erinnerungen*, 51–52).

[80] Ibid., 57.

[81] Ibid., 63.

[82] Ibid., 69.

[83] Johann Friedrich A. Göttling, "Einige Bemerkungen über Chemie und Pharmazie in England," *Almanach* 1789, 120–44.

[84] See also Hufbauer, *German Chemical Community*, 208.

[85] Göttling, "England," 124.

[86] Ibid., 127.

[87] Ibid., 128–44.

[88] Ibid., 122.

[89] On the school, see above. On Neumann's professional career see Alfred Exner, *Der Hofapotheker Caspar Neumann (1683–1737)* (Berlin: Triltsch & Huther, 1938).

[90] See Exner, *Neumann*, 9.

91 Exner emphasized that at the time "chemistry was mainly cultivated in Germany in mines and smelting works" (Exner, *Neumann*, 9). This may be an overstatement, but it is important to note that in eighteenth-century Germany, as in Sweden, metallurgy was a very significant site of chemical practice.

92 See Exner, *Neumann*, 25. Observation of facts in the arts and craft was a part of "experimental history" (*historia experimentalis*) in eighteenth-century chemistry. On *historia experimentalis* see Ursula Klein, "Experimental History and Herman Boerhaave's Chemistry of Plants," *Studies in History and Philosophy of Biological and Biomedical Sciences* 34, 2003, 533–67 and "Experiments at the Intersection of Experimental History, Technological Inquiry, and Conceptually Driven Analysis: A Case Study from Early Nineteenth-Century France," *Perspectives on Science* 13, 2005, 1–48; Ursula Klein and Wolfgang Lefèvre, *Materials in Eighteenth-Century Science: A Historical Ontology* (Cambridge, MA: MIT Press, 2007).

93 See Exner, *Neumann*; Grünhagen, "Einrichtung Apotheken"; Manfred Stürzbecher, *Berlins alte Apotheken* (Berlin: Verlag Bruno Hessling, 1965).

94 Hufbauer, *German Chemical Community*, 62.

95 William H. Brock, *The Norton History of Chemistry* (New York: W. W. Norton, 1993), 436.

96 This attitude accorded with the venture of natural and experimental history of the time (see footnote 92).

97 Lorenz Crell, "Vorrede," *Chemisches Journal* 1778, 9–20, on 12. In 1781 the journal was renamed *Die neuesten Entdeckungen in der Chemie*, and in 1784 it was renamed again to *Chemische Annalen für die Freunde der Naturlehre, Arzneygelahrtheit, Haushaltungskunst und Manufakturen*. For a more detailed analysis of Crell's journal see Hufbauer, *German Chemical Community*, 62–95.

98 Crell, "Vorrede," 19.

99 The list is preserved almost in its entirety, but Königsberg is missing. It provides information about the names of the owners of apothecary's shops, administrators (*Provisor*) and clerks, and of the town in which the apothecary's shops were located. In 1928 Adlung transcribed and published the list in alphabetical order by name; see Alfred Adlung, "Apothekenbesitzer, Apothekengehilfen und –lehrlinge Preußens im Jahre 1798," *Archiv für Sippenforschung und alle verwandten Gebiete* 5, 1928, 163–66, 200–03; 229–32, 280–83. See also Alfred Adlung, "Alte Apothekerfamilien und ihre Apotheken," *Pharmazeutische Zeitung* 73, 1928, 1453–60.

100 As information about subscribers exists only for the earlier period of 1784–89, a direct, accurate comparison of the individual subscribers and the total number of Prussian apothecaries (owners and *Provisors* of apothecary's shops) is not possible. Hence my comparison yields only an estimation, which relies on the presupposition that the number of apothecaries did not change dramatically in the period from 1784 to 1800. For a list of all subscribers to Crell's journal see Hufbauer, "German Chemical Community," 272–99.

101 Nevertheless they were distributed over the whole country, including such distant towns as Bielefeld, Halle and Stettin.

102 See my tentative distinction of four different groups of apothecaries above.

103 For the impact on Crell's income see Hufbauer, *German Chemical Community*, 85.

104 Martius, *Erinnerungen*, 63.

105 Ibid., 39.

106 The name *Almanach* (almanac) refers to a calendar included at the beginning of the journal, which listed pharmaceutical preparations for each month.

107 Johann Friedrich A. Göttling, "Vorbericht," *Almanach* 1786. On Göttling's journal see also Bettina Wahrig, "Apotheke – Öffentlichkeit – Publikum. Zur Geschichte eines Dreiecksverhältnisses," 9–28, in *Apotheke und Publikum: Die Vorträge der Pharmaziehistorischen Biennale in Karlsruhe vom 26. bis 28. April 2002*, Christoph Friedrich and Wolf-Dieter Müller-Jahncke, eds. (Stuttgart: Wissenschaftliche Verlagsgesellschaft, 2003).

108 Göttling, "Vorbericht," *Almanach*, 1792.

109 Ibid. In the subsequent volume Göttling wrote that his "pocketbook is mainly destined to seek out the defects of the apothecary trade, to make proposals for their improvement, and to banish them," "Vorerinnerung," *Almanach* 1793.

110 Göttling, "Vorbericht," *Almanach* 1783.

111 See footnote 92.

112 Göttling, "Vorbericht," *Almanach* 1785.

[113] Lorenz Crell, "Anzeige chemischer Schriften: Almanach oder Taschenbuch für Scheidekünstler und Apotheker auf das Jahr 1781," *Die neuesten Entdeckungen in der Chemie* 1, 233–34. Unfortunately, Göttling did not provide information about the names or number of subscribers to his journal.

[114] Ibid., 234.

[115] Göttling, "Vorbericht," *Almanach* 1787.

[116] Lorenz Crell, "Anzeige chemischer Schriften: Almanach oder Taschenbuch für Scheidekünstler und Apotheker auf das J. 1790. Erstes Jahr," *Chemische Annalen*, pt. 2, 1790, 79–85, on 79.

[117] Göttling, *Almanach* 1789, 176–81, on 176–77.

[118] For further details, see Klein, "Innovation."

[119] Hufbauer, *German Chemical Community*, 62 and 89.

[120] See also Klein, "Experiments at the Intersection," and Klein and Lefèvre, *Materials*.

[121] See Hufbauer, *German Chemical Community*, 180–81; Stürzbecher, "Apotheken," 44–45; Georg Edmund Dann, "Marggraf Briefe," *Geschichte der Pharmazie* 20, 1968, 20–23.

[122] The papers were published in the *Miscellanea Berolinensia* (in Latin) and the *Histoire de l'Académie Royale des Sciences et des Belles-Lettres de Berlin* (in French) of the Berlin Society of Sciences and later, Academy of Sciences, respectively (from 1744). A list of Marggraf's works is contained in: Otto Köhnke, *Gesammtregister über die in den Schriften der Akademie von 1700–1899 erschienenen wissenschaftlichen Abhandlungen und Festreden* (Berlin: Reichsdruckerei, 1900). Marggraf's papers were also republished in German: Andreas Sigismund Marggraf, *Chymische Schriften*, 2 vols. (Berlin, 1761–67).

[123] See "Archiv der Berlin-Brandenburgischen Akademie der Wissenschaften, Bestand Preußische Akademie der Wissenschaften (1700–1811)": I–XIII–19, folio 36–47; in the following abbreviated as ABBAW.

[124] Lorenz Crell, "Lebensgeschichte Andreas Sigismund Marggraf's, Directors der physikalischen Klasse der Königl. Preuß. Akademie, Mitglieds der K. Akademie der Wissenschaften zu Paris und der Churfürstl. Maynzischen zu Erfurt," *Chemische Annalen*, pt. 1, 1786, 181–92, on 181.

[125] Crell, "Lebensgeschichte Marggraf's," 182–83.

[126] For Klaproth's biography, see Dann, *Klaproth* and "Klaproth, Paecken und Liphardt," *Pharmazeutische Zeitung* 26, 1971, 932–39); E. G. Fischer, "Denkschrift auf Klaproth," *Abhandlungen der Königlichen Akademie der Wissenschaften in Berlin* 1818/19, 1820, 11–26; Michael Engel, "Zwischen Beruf und Berufung: Klaproth, Hermbstädt, Rose, Gehlen – Aus dem chemischen Alltag zwischen 1775 und 1800 am Beispiel Berlin," 259–83 in *Apotheker und Universität: die Vorträge der Pharmaziehistorichen Biennale in Leipzig vom 12. bis 14. Mai 2000*, Christoph Friedrich and Wolf-Dieter Müller-Jahncke, eds. (Stuttgart: Wissenschaftliche Verlagsgesellschaft, 2002).

[127] The description was part of Johann Friedrich Westrumb's autobiographical notes, which were published by August Westrumb, "Johann Friedrich Westrumb, Dr. der Medicin, Königl. Großbrit. Hannover. Berg-Commissar und Apotheker in Hameln," *Neues Vaterländisches Archiv, Hannover* 7, 1825, 23–42, 26. It should be noted that Westrumb's MD was a honorary degree from Marburg University, awarded in 1811.

[128] Mohrenstraße 5 (Zietenplatz) is only a few meters from the Czech Embassy that hosted the Max Planck Institute for the History of Science from 1994 until 2006.

[129] He also enjoyed the tutelage of Rose's children, one of whom, Valentin Rose Jr. (1762–1807), became a well-known apothecary-chemist.

[130] See Klaproth's list of publications included in Dann, *Klaproth*, 110–28.

[131] For a scientific biography of Achard see Hans-Heinrich Müller, *Franz Carl Achard (1753 bis 1821)*, (Berlin: Verlag Dr. Albert Bartens, 2002).

[132] See ABBAW I–XIII–26, folio 11.

[133] See ABBAW I–XIII–26, folio 131.

[134] See ABBAW I–XIII–26, folio 65–66.

[135] Hufbauer presented the pharmaceutical art as a "realm of recipes," and the apothecaries who also became chemists as men who "transcended" this realm ("German Chemical Community," 55). Accordingly, he asserted that there was "a gap between compounding medicines and pursuing chemistry" (56). More recently, Simon made similar claims with respect to late eighteenth-century French pharmacy (Simon, "Chemistry").

[136] On cameralism and chemically trained cameralists in eighteenth-century Germany, see R. Andre Wakefield, "Police Chemistry," *Science in Context* 12, 1999, 231–67.

[137] See Simon Schaffer, "Experimenter's Techniques, Dyer's Hands, and the Electrical Planetarium," *Isis* 88, 1997, 456–83, on 482.

[138] See Steven Shapin, *A Social History of Truth* (Chicago: University of Chicago Press, 1994).

[139] As I did not scrutinize in this paper the contents of the chemical texts written by apothecary-chemists, I omit here epistemological criteria. Parenthetically, however, I would like to add that there is no way to sort out the books and journal articles on chemistry written by well-known German apothecary-chemists (my group four) and chemists educated at universities; see Klein, "Innovation."

[140] This thesis is usually ascribed to Edgar Zilsel, but I found no unambiguous evidence that Zilsel actually postulated a hybrid persona of scholar and craftsman in the early modern period. What he claimed repeatedly was that the social prejudice against manual labor was overcome, and that in so doing rationally trained scholars adopted the experimental method. See, for example, Edgar Zilsel, "The Sociological Roots of Science," *American Journal of Sociology* 47, 1942, 544–62.

[141] Rupert Hall, "The Scholar and the Craftsman in the Scientific Revolution," 3–23 in *Critical Problems in the History of Science: Proceedings of the Institute for the History of Science at the University of Wisconsin, September 11, 1957*, Marshall Clagett, ed. (Madison: University of Wisconsin Press, 1959), 17.

[142] Hall, "Scholar and Craftsman," 7.

[143] Thomas S. Kuhn, "Mathematical versus Experimental Traditions in the Development of Physical Science," 31–65 in *The Essential Tension: Selected Studies in Scientific Tradition and Change*, Thomas S. Kuhn, ed. (Chicago: The University of Chicago Press, 1977). Kuhn attributed some stronger influence by practitioners on the transformations of the classical sciences than Hall; for example, he also stated that "the concern of artist-engineers with these classical fields was a significant factor in their transformation," 56.

[144] It is questionable whether Hall's and Kuhn's views were correct with respect to other sciences, but I do not deal with this issue here.

[145] Hall, "Scholar and Craftsman," 19.

[146] Ibid.; weighing was indeed an intrinsic concern of practioners such as apothecaries and assayers. For the relationship between assayers and chemists in the early modern period, see William R. Newman and Lawrence M. Principe, *Alchemy Tried in the Fire: Starkey, Boyle, and the Fate of Helmontian Chymistry* (Chicago: The University of Chicago Press, 2002); Ursula Klein, *Verbindung und Affinitität: die Grundlegung der Neuzeitlichen Chemie an der Wende vom 17. zum 18. Jahrhundert* (Basel: Birkhäuser, 1994).

[147] Kuhn, "Traditions," 57.

[148] Ibid, 58: "Excepting chemistry, which had found a variegated institutional base by the end of the seventeenth century, the Baconian and classical sciences flourished in different national settings from at least 1700."

[149] Especially in Germany, the teaching of chemistry at medical faculties had a long tradition going back to the first decades of the seventeenth century. Between 1720 and 1780 the number of salaried chairs for chemistry at German medical faculties rose from six in 1720 to twenty-eight in 1780; see Hufbauer, *German Chemical Community*, 34.

[150] It should be noted that on the eighteenth-century European continent the term "laboratory" referred mostly to sites of pharmaceutical manufacture and chemical experimentation, as well as to places where production and chemical technological inquiry intersected, such as arsenals, assaying shops, distilleries and mining boards. By contrast, eighteenth-century experimental philosophy or physics, as a rule, did not establish laboratories.

MATTHEW D. EDDY

THE ABERDEEN AGRICOLA: CHEMICAL PRINCIPLES AND PRACTICE IN JAMES ANDERSON'S GEORGICS AND GEOLOGY

INTRODUCTION

Late eighteenth-century industrialists, farmers and physicians who actively employed chemistry are historically hazy figures. This is not only the case for scientific histories, but also for studies that address the socio-economic factors of the Enlightenment. Up until the late twentieth century, historians of chemistry tended to focus on ideas and personalities that nineteenth-century scholars deemed to be important to Antoine Lavoisier's oxygen theory of combustion – a practice that fell in line with the more ubiquitous 'Great Man' approach to history. Such a move privileged the idealized space of the laboratory, thereby ruling out experiments performed *in situ* in homes, farms, mines, factories, or fields. The result was a historiographical disposition that assigned a causative role to the chemical concepts developed in conjunction with the instruments and methods attributed to laboratory settings: an act that reconfirmed the status of canonized chemists, and implicitly made them central nodes in knowledge networks that honed and dispersed the theoretical framework of chemistry. Although this model was conceptually useful (especially in light of the accessibility of primary sources), it effectively marginalized provincial chemists (among others) by placing them at the end of a long chain of ideas that emanated from a distant expert. Such a one-directional approach treated local chemists as if they were intellectual automatons waiting to be directed by an unseen hand.

From the 1970s onward, the Lavoisierian historiography slowly began to metamorphosize; initially through the work of Reijer Hookyaas and Frederic L. Holmes, and then by a number of other scholars. Within these studies, it was emphasized that there were national contexts of chemistry that were influenced both by language and other localized intellectual predispositions. Germany, France, and Sweden received their due attention, as well as Italy, England, and Scotland.[1] However, once again, the role of local actors outside the traditional laboratory setting in these studies was predominantly one of passivity. A very good example of this state of affairs can be seen in the case of Scotland. Over the years, attention was devoted to William Cullen's research on heat and Joseph Black's work on fixed air (carbon dioxide), but the local specifics of their interlocutors and the application of their ideas to industry and georgics (farming) have remained relatively unexamined.[2] This is in spite of the

L. M. Principe (ed.), New Narratives in Eighteenth-Century Chemistry: Contributions from the First Francis Bacon Workshop, 21–23 April 2005, 139–156.

fact that Scotland's highly literate Lowlands population harbored a high number of university trained chemists. The purpose of this paper is, therefore, to highlight the career and thought of one such person. Although there are many notable candidates who have received little attention (Dr. John Roebuck and Rev. Dr. William Laing for instance),[3] I have chosen to focus this essay on Dr. James Anderson (1740–1809).

Anderson wrote on many chemical topics, but I will focus specifically on his understanding and use of a substance that he called lime. After a few initial comments on his education and vocation as a publisher, I set out the chemical principles taught to him by Cullen that framed his understanding of properties of this substance. Focusing on his popular 1775 "Essay on Quicklime," I detail the lime experiments that he conducted on his own estate. Since the aim of his essay was to show farmers how to improve the economic productivity of their land, he skillfully broke down the concepts of Edinburgh's chemists into clear, succinct terms and he used "instruments" that could be obtained on just about any farm. In addition to chemistry's georgical utility, Anderson also made a significant connection between his brand of *in situ* analysis and the composition of local minerals. More specifically, he used his experimental knowledge of cements to postulate the age of limestone strata that he encountered in both England and Scotland. This, I suggest, is a notable (and potentially common) example of provincial ingenuity that fused the principles and practices of chemistry into a coherent explanation of a naturally occurring formation that was relevant to the nascent field of geology.

THE PRACTICAL PUBLISHER

Over the past decade there has been a renewed interest in the Scottish Enlightenment that has engendered the publication of many books that seek to explain how the Scots saved late eighteenth-century Britain from intellectual and economic torpor. Like a tale twice-told, the cast of these books, though dressed in different garments, inevitably remains the same. In the end, one is often led to think that David Hume, Adam Smith, Thomas Reid, Dugald Stewart and the like were really the only serious thinkers of the time. However, one look at the archives housed in just about any Scottish university reveals that this is not the case. The natural philosophy that these intellectuals espoused was challenged and augmented by many men and women whose names are now seldom mentioned. One of these people was Dr. James Anderson.

James Anderson was born outside of Edinburgh in 1740 and died in Essex in 1809.[4] As a teenager he inherited his parents' farm after their untimely deaths, and in 1768 he became the owner of a large estate when he married Margaret Seton, a wealthy Aberdeen heiress. Like most farmers, he was keenly interested in improving the economic potential of his land. He published many works on this subject, often under the simple pseudonym of "Agricola" – a name which gestured towards the influential writings of the natural historian Georg Agricola (1494–1555). Anderson soon realized that the pen was just as lucrative as the plough and began to write newspaper and magazine articles. This proved to be financially rewarding and so during the 1770s he decided to try his hand at authoring pamphlets and books. Several of his earlier works were politically charged, but the bulk of his writing addressed the industrial and agricultural improvement of Scotland.[5] His first major work in this area was *Essays Relating to Agriculture and Rural Affairs* (1775). It sold well; over the next 20 years he went on to expand it into several volumes. This success led him to pen works on the related topics of industry, forestry, fishing, economics, alternative fuels, and coal mining.[6] In the 1790s he moved into publishing periodicals, first *The Bee* and then *Recreations in Agriculture, Natural-History, Arts and Miscellaneous Literature.*[7]

Though historians have concentrated primarily on the political economy his writings advocate, it is seldom noticed that much of his agricultural thinking was based on chemistry. The wider context in which chemistry was used in this manner in Enlightenment Scotland has been addressed by several authors,[8] but to set the stage for Anderson's perception of lime it is necessary to mention a few key points. As a young man he attended the University of Edinburgh where he was particularly taken by the chemistry lectures of Professor William Cullen. Throughout the course of his career, he endeavored to apply the chemical theory of Cullen's lectures to the practical concerns of productivity that he encountered on his farms and while travelling around Scotland. One of key practices that Anderson took from Cullen's lectures was the value of *in situ* chemical analysis. Almost all of the experiments that he recounts in his works were conducted either in the field or on his own farm. He often included his results in his publications, and over the course of his career he became respected for his knowledge of chemistry. Indeed, he spoke on chemical matters to the Royal Society of Edinburgh[9] and one of his articles in *The Bee* is cited in Samuel Parkes' popular early nineteenth-century textbook *A Chemical Catechism.*[10]

Anderson's chemical understanding of lime is best evinced by the lengthy "Essay on Quicklime" that he published in the aforementioned *Essays Relating to Agriculture and Rural Affairs*. The entire work was dedicated to William Cullen and, with tear-jerking rhetorical flair, Anderson stated that Cullen had guided him when no other person had been there to direct his youthful steps. Unfortunately, in the years that followed, Cullen's son, Judge Advocate Robert Cullen (Lord Cullen), did not share this sentiment and sued Anderson for plagiarism. Anderson won the case, but his pride was irreparably wounded.[11] Despite this legal battle, *Essays Relating to Agriculture* went through at least four editions over the next three decades. Since the work was so popular, examining its content can help shed light on the type of chemistry being read by gentlemen farmers in Enlightenment Scotland; thereby moving the focus from the professors in the University of Edinburgh to their students throughout the city and the countryside.

THE CHEMISTRY OF LIME

During the 1750s, the professors of the University of Edinburgh held that the most basic forms of matter were water, earths, salts, inflammables, and metals.[12] During the early 1760s, Joseph Black tentatively added another: air (or aer).[13] In Edinburgh, the chemistry lectures of Cullen (1755–1766) and Black (1766–1799) used the terms "principles"[14] and "elements"[15] interchangeably to describe these forms. In this essay,

I will use the term "principle" to avoid any confusion. The above principles functioned as the building blocks of all quantifiable material substances, bodies, and part(icle)s able to be manipulated in the laboratory.[16] Likewise, a "compound" was the combination of any principle with another principle, or with another compound. Attraction between the principles was based on the affinity tables proposed by Etienne-François Geoffroy (1672–1731).[17] Based on this system, quicklime proved to be a very tricky substance to classify because it contained "lime." Some of Edinburgh's chemists argued that lime was either an earth or a salt, while others suggested that it might be a compound of earth and salt. In offering such different definitions, they followed in the footsteps of many natural philosophers from antiquity through to early modernity who struggled to understand the substance. In the words of Robert Multhauf, "Quicklime was as recalcitrant as sulphur to scientific elucidation."[18]

During the 1740s, quicklime gained the attention of Edinburgh's chemists on account of its relation to limewater, a therapeutic cure made from limestone that was used to remove bladder stones (*calculi*).[19] As their names indicate, the common ingredient found in quicklime, limestone, and limewater was lime. This substance expressed different qualities when exposed to fire, water, or salts (acids and alkalis). If one heated limestone to an extreme temperature, it would crumble into quicklime, a chalky dust. If water was poured on the quicklime, it would settle into two layers, one of paste on the bottom and one of liquid on the top. This top layer of liquid was called limewater. During the 1740s and 1750s there was a series of intense debates between professors Charles Alston (1685–1760) and Robert Whytt (1714–1766) over the chemical composition and therapeutic value of the lime in the water.[20] This topic proved to be a conceptual touchstone for the medical faculty and was therefore taken up by the chemist William Cullen when he transferred to Edinburgh from Glasgow in 1755.

Although Alston and Whytt disagreed about the specific medical use of lime, they both agreed that it was the key ingredient of limestone. Based on Edinburgh's fascination for experiments with lime, and on his training under Cullen, Joseph Black, Cullen's star student, took this debate to another level with his 1754 medical thesis entitled "On the acid humour arising from foods, and on magnesia alba."[21] The work received such a favorable response that he expanded it to cover quicklime and then published it in the 1756 edition of Edinburgh's international journal *Essays and Observations*.[22] Black's work recounted how he had used heat and acids to break down quicklime into its most reducible components. Based on gravimetric analysis, his results allowed him to argue that naturally occurring limestone essentially consisted of "Calcareous Earth," a compound in which lime was intimately united with fixed air. Over the next decade, he performed numerous experiments that allowed him to expand his fixed air thesis so that by the 1760s Edinburgh chemists trained by him believed that limestone was held together by an affinity that existed between lime and fixed air.[23]

Black's experiments turned out to be applicable to wide number of chemical practices that produced economic rewards. For industry, his interpretation of lime was applied to the manufacture of linen.[24] For medicine, his essay inspired experiments that sought to determine whether limewater could be used to dislodge bladder stones or to combat upset stomachs. Several useful cures were developed based on Black's work and their efficacy quickly guaranteed his essay's central importance for medical therapeutics. Because of this success, it was not long before Black's work was applied to topics that fell outside the direct remit of medicine. In particular, it was used to expand the chemical utility of quicklime, a substance that had long played an important role in agriculture and which was being used more frequently in the chemical processes of Scotland's early industrial revolution. This new application was evinced in a number of medical and industrial publications that appeared during the 1760s and 1770s. For example, there was David MacBride's *Experimental Essays on Medical and Philosophical Subjects*,[25] which contained a section on "The Dissolvent Power of Quicklime," and there was also Francis Home's compilation *Experiments on Bleaching*, which included essays written by Black, Home and MacBride.[26] Thus, when Anderson's "Essay on Quicklime" first appeared in 1775, it was following a larger trend in which the University's experiments were being used outside the laboratory to advance Scotland's economic potential.[27]

PRINCIPLES OF COMPOSITION

In the first sentence of the "Advertisement" prefixed to Anderson's "Essay on Quicklime," he states that "The nature of the subject ... should be treated in a more scientific manner." By "scientific," he meant *quantitative* and the best way to gather this sort of data in Edinburgh was via chemistry. Anderson's chemistry fell firmly within the chemical nomenclature advocated by Cullen and Black. Like so many students who had studied chemistry in Edinburgh from the 1750s to the 1790s, Anderson was not eager to convert his system's terms and definitions into those proposed in Lavoisier's 1787 nomenclature. As he stated in a footnote in the 1797 edition: "Since this Essay was written, a total change has taken place in regard to the names of chemical substances – but I do not think it necessary here to make any change in that respect, the terms being here all explained as they occur, so as to prevent ambiguity."[28] As other comments throughout the 1797 version of the quicklime essay and his other works indicate, Anderson was indeed familiar with the new nomenclature. However, even though Lavoisier's oxygen theory had gained a noticeable following among several of the leading chemists of University of Edinburgh during the 1790s, its uptake among practicing physicians, apothecaries, and gentleman farmers in Scotland was much slower.[29] It was not until around 1800 that the new nomenclature filtered into all of the medical school's courses and into more popular publications like the genteel *Scots Magazine*.

The principle-based matter theory that Anderson offered his readers was a simplified, clearly stated version of the chemistry of lime as presented in the lectures of Cullen and Black. More specifically, Anderson concentrated strictly on the chemical composition of quicklime. To make the essay easier to understand for those not trained in chemistry, he took care to offer footnotes that gave the definitions of chemical terms: "To avoid disagreeable circumlocution, I shall be obliged, in this Essay, to employ some technical terms not commonly understood; but shall explain their meaning as I go along."[30] Following this method, he defined around thirty key substances, processes, or states to provide the reader with the basic vocabulary necessary to understand the composition of quicklime. In this sense, the essay was a focused summary and application of Black's experiments.[31] Though the essay does briefly mention the principles of air, fire, and water, most of it focuses more on earths and salts. For those more familiar with chemistry, Anderson included a detailed postscript entitled "Directions for ascertaining the purity of Lime, and discovering the Nature of the Bodies mixed with it." Additionally, since Scotland and England disagreed over a common unit of mass during the late eighteenth century, Anderson avoided specific weights and measures by using ratios of composition.[32]

Following Black, Anderson held that pure limestone was made up of calcareous earth united with fixed air. In his words, calcareous earth was "a general term denoting all those substances that consist of the matter which lime may be made, in whatever state it may be found – whether alone – or mixed with other substances, that prevent it from being reduced to powder after calcination."[33] For reasons of simplicity, he often used "lime" synonymously with "calcareous earth."[34] Lime, as he explained, existed in three states. First, "mild" calcareous earth was limestone in its pre-calcined state. Second, "Caustic" calcareous earth was "exactly synonymous with *quicklime*, in its strict and philosophical acceptation."[35] Finally, "Effete" calcareous earth was the hardened cement that formed from the quicklime; which meant that it was a post-calcination form of limestone. As Anderson explained it: "Lime is no sooner slaked, than it immediately begins to absorb its air, and return to its former *mild* state; or, in other words, it becomes *effete*; in which state it possesses the same chemical qualities, in every respect, as limestone."[36]

These different states were based on Black's 1756 article in *Essays and Reviews*. To help his reader, Anderson offered his own nine variety arrangement of mild calcareous earth (since the caustic and effete were not stable).[37] For each variety, he took care to note two classification characters: "purity" and "hardness." Purity referred to the percentage of mild calcareous earth contained in the stones, something that could be visually inferred from the stone's transparency. Hardness referred to how much the stone had been able to crystallize – a process in which "saline substances, when dissolved in water, and put into proper circumstances for that purpose, separate from the water, and *shoot* into regular figures, which assume different forms, and are more or less transparent according to the different nature of the salt, as nitre, alum, &c."[38] After setting out this classification, Anderson then proceeded to explain how chemistry could be used to find and assay calcareous minerals.

Such a classificatory approach to limestone (a mineral) effectively mirrored the methods used in the chemical mineralogy that was being taught and disseminated at the University of Edinburgh. During the 1750s, Edinburgh's professors and their students that were sent out to London and the colonies were quite interested in the principles of earth and salt.[39] Based on his experiments, Cullen held that a "Calcareous Earth" had two distinguishing properties: alkalinity and effervescence with acids.[40] Since pulverized (mild) calcareous earth dissolved in water, but could harden also into non-crystallized stone, Cullen was quite torn as to whether it was an earth or a salt.[41] Anderson's "Essay on Quicklime" shows that he too struggled with this problem. Although he classified calcareous "matter" as an earth (above), he called it "saline" in other parts of the essay.[42] This was relatively the same approach followed by Black and John Walker, two of Cullen's students. Walker had published on salts in the *Philosophical Transactions* in 1757[43] and Black's 1750s experiments, as I mentioned above, had led him to conclude that calcareous earth was a *compound* made up of lime (that had saline properties) and fixed air. Even so, they both followed Cullen's lead and classified calcareous matter as an "Earth." Walker made it a genus, while Black made it a species of "Absorbent Earth." This being the case, Anderson's classification of calcareous earths followed Black,[44] whereas Walker's, who accepted Black's fixed air thesis, did not.[45] Black was Professor of Chemistry from 1766 to 1799 and Walker was Professor of Natural History (which included mineralogy and geology) from 1779 until 1803. Neither Black's genus for "Absorbent Earth,"[46] nor Walker's for "Calcareous Earth,"[47] changed during their tenure, and Anderson's "Essay on Quicklime" worked well with both of their arrangements.

FROM PRINCIPLES TO PRACTICE

The main goal of Anderson's essay was to extend analytic "rules" into georgics and to introduce chemical analysis as an agricultural tool.[48] In principle, these two objectives could be treated separately, but in practice Anderson used chemical data to promote an analytical approach to mineralogical utility and composition. To dig into the specifics of limestone composition, Anderson offered several chemical processes to help his readers design experiments that could separate a stone's calcareous earth from other substances like siliceous earth, argillaceous earth, animal gluten, and salts. For analysis, he describes how to use fire (calcination and vitrification) and solvents (water and salts), which he usually called menstrua. These types of processes were known not only to Anderson, but also to those who had studied chemistry in Scottish universities and to those who were apothecaries or who worked in industrial settings. This cast of chemists was encouraged to use fire and menstrua on an equal basis, but, in practice, menstrua were used more often because attaining high heat outside the laboratory was often difficult. Following a practice that was commonplace in early modern chemical mineralogy, Anderson encouraged farmers to use these processes to gravimetrically track each chemical component of a limestone specimen. Once the weight of a component was determined, it was divided by the total weight of the original specimen to produce a ratio of composition.[49] In the end, a higher the ratio of pure calcareous earth meant that the specimen was worth more money. It was therefore to the farmer's advantage to use chemistry to determine the specific composition of the limestone on his own land or in the specimens that he was buying to make cement or manure.

Although many of the mineralogical assaying methods developed by the medical school were quite complex, Anderson was able to encourage gentleman farmers to become chemical philosophers because he had picked a relatively simple analytical procedure that only used menstrua as solvents (that is, fire was not needed). He summarized the process in the following manner: "As all calcareous matters are capable of being dissolved in acids ... it follows, that if a sufficient quantity of acid is poured upon any body that contains calcareous matter, this matter will be quickly dissolved, while the others are left behind; and the proportions of each may be accurately ascertained."[50]

Over the course of the eighteenth century, the instruments used in the chemistry laboratories of Edinburgh became more elaborate and complex. However, for Anderson's experiments, they were actually quite simple. They consisted of a glass or earthen vessel, a tobacco pipe bit (for stirring), filter paper, and a glass funnel. All of these items could be found lying around the average farm, save for the glass funnel, which could have been obtained from an apothecary. The only chemical needed to perform the experiment was a strong acid. Anderson explained that *aquafortis* (nitric acid) was the best one for the job, but that spirit of salt (hydrochloric acid) would do in a pinch.[51] Obtaining *aquafortis*, as Anderson notes, was not difficult because it could be "bought in the shops, for the use of dyers."[52] Other acids were also readily available in the Lowlands because they were used by physicians, bleachers, potters, apothecaries, and saddlers or they were produced on industrial sites.[53] In fact, Anderson thought acids to be so helpful for agricultural chemistry that he encouraged all country gentlemen "to keep a phial of aquafortis, or muriatic acid, always with him, for making trials of calcareous substances: the expense is nothing; and I am persuaded, from the want of it alone, many persons have failed to make discoveries of calcareous matters."[54]

Yet the chemically savvy farmer could not possibly assay every piece of limestone that was laid in front of him. It is here where Anderson's use of compositional ratios gives way to probability. He argued that anyone who wished to make an accurate analysis of any newly discovered limestone, "will do well to take eight or ten stones from different parts of the quarry, that are somewhat different in appearance from one another; and, having taken a chip from each, pound the whole together, to afford a proper subject for the experiment."[55] Using this approach, a farmer could employ chemistry to predict the quantity of calcareous earth contained in local varieties of stones. Once this amount was determined, the farmer could then use the purity of his estate's limestone to estimate whether it was cheaper to import quicklime or to make it himself. Such an approach was not only more methodologically exact, but it was also more lucrative – a point well worth noting when one considers that the Lowlands were rich in different varieties of limestone. Additionally, Anderson suggested that a farmer could calculate the probable yield of a field by comparing its annual bushel production to the amount of calcareous earth contained in the soil.[56] By advocating this combination of probability and chemistry, Anderson held that he was providing farmers with a method and "facts" that would allow them objectively to improve the economic productions of their estates. In so doing, he was following a larger movement in which chemical practices developed in the laboratories of the University of Edinburgh were placed in conversation with industrial, agricultural, and medical experimentation.[57]

In applying chemistry to agriculture, Anderson was following an established tradition in Edinburgh. Based on his experience on his farm in Ormiston, Cullen had previously given a series of lectures on agriculture in the 1740s and 1750s,[58] and then another in the 1760s[59] that was eventually published posthumously.[60] Since agriculture was central to Scotland's economy, lecturing on georgics was a lucrative pursuit, especially if given to a private audience. Even Black and Walker frequently made asides in their lectures that were directly relevant to agricultural topics like manures, cements, and topsoil.[61] Black assayed earths sent to him by farmers[62] and commented on plant nutrition, and Walker gave an entire series of georgic lectures in an (unsuccessful) attempt to be appointed to the Chair of Agriculture during the 1790s.[63] Such an academic interest in the chemistry of georgics also engendered a limited number of specialized papers given in the societies of Edinburgh, Aberdeen, and Glasgow and a smaller, but notable, corpus of specialized essays or medical dissertations that were published either individually or as compilations. Anderson's work extended and augmented this tradition and, as a result, his *Essays* went through four printings.

FROM PRACTICE TO PRINCIPLES

Anderson's chemical understanding of lime and its various material states directly impacted his view of the form and composition of the Scottish landscape, and this in turn affected his thoughts on the nascent field of geology. At this time in Edinburgh geology was the domain of the medical school; it was one of six topics taught in Walker's year-long natural history course. Relevant mineralogical topics were also covered throughout Black's chemistry lectures. Both followed the predominant early modern notion that there were three types of geological strata: primary, secondary, and tertiary.[64] Primary strata were the bottommost layer and consisted of hard rocks like granite and jasper. Secondary strata contained softer rocks like slate, marble, and limestone. Tertiary strata were mostly soil and deposits made by rivers. Like most chemically-trained natural historians, Anderson viewed the composition of strata, especially limestone, via the experiments that he had conducted on limewater and quicklime. Since the context of his chemical experimentation was closely based upon his agricultural experiments, his knowledge of geological strata was heavily informed by practical application.

Anderson's views on secondary strata are most clearly expressed in sections 10 and 11 of his "Essay on Quicklime" where he addressed the formation of limy stalactites.[65] Caves were a very popular attraction in Scotland at this time, especially Fingal's Cave in the Hebrides. Anderson explained that the original "nipple" of a stalactite was formed by "natural limewater" (caustic calcareous earth) that ran across the top of a cave. As the drop moved, air ran across its surface thereby forming a slightly solidified outer shell (mild calcareous earth) that then stuck to the top of the cave. From this formed a "pendent concretion, resembling the shape of an icicle" (an object that Anderson called a "tangle")[66] that grew downward to the cave's floor and where it then formed a large sheet of limestone. It is here where his agricultural chemistry connects with geology, because Anderson argues that the formation of the stalactite's sheet-like base could be used to understand the formation of limestone or marble strata. He held that if there was a considerable stream of limey water, these types of strata could form in a relatively short period of time. More specifically, he states that, "In this manner, within the memory of man, have huge rocks of marble been formed near Matlock, in Derbyshire." He qualifies this by averring that other forms of limestone might take more time to form, but his use of limey stalactites to explain the accretion of geological strata is clear, as he wrote, "But there is no room to doubt, that all the strata of calcareous matter in the world, have been formed by a process exactly similar to this."[67]

The "Essay on Quicklime" also contains important points that were relevant to eighteenth-century perceptions of primary strata. These views were directly linked to Anderson's understanding of crystals and cementation as described in the essay. In his discussion of limestone "purity," he included a footnote on crystal formation that ran for three pages. There he reminds his readers that crystals were formed from salts and that they "contain a considerable proportion of water united with the saline matter."[68] The more plentiful the crystallization, the harder the limestone. Later in the essay he picks up on this point and states that lime becomes cement via a "certain degree" of crystallization. He added that if sand granules were mixed into the quicklime, they would stimulate the growth of stronger crystals and, hence, create a harder cemented mass. As he described it:

> ... the particles of hard sand, like sticks or threads, when making sugar candy and other chrystals, while surrounded by the watery solution, will help to forward the chrystallization, and render it more perfect [harder] than it otherwise would have been, so as firmly to cement the particles of sand to one another.[69]

This link between "chrystallization" and the concretions formed from quicklime is an important point because Anderson's Edinburgh contemporaries held that geological strata were held together by "cement."[70] Walker, for instance, held that natural occurring cement held most strata together. Furthermore, Anderson held that lime that had cemented (hardened) underwater was much firmer than that hardened above the surface of water. His knowledge of this subject stemmed from his writing about underwater cementation in other works that addressed dams, bogs, and bulwarks.[71] The need for water to form a hard cement, like that which held together primary strata, suggests that Anderson, like many of his contemporaries including Walker and Black,[72] held that all observable strata were formed via an aqueous medium that had at one time flooded the earth.[73]

Anderson was able to move from chemical "facts" to geological "phenomena" because he was using the same axiomatic method that allowed him to extend the rules of chemistry into agriculture. In his discussion of stalactites and limestone strata, he differentiated between the "Operations of Art" and the "Operations of Nature." The former were, in this case, the principles of chemistry.[74] He used these to obtain the "facts" of composition and then he noted other properties like shape and mass so that he could then inductively determine the process by which the stalactite had formed. Once this was done, he could then address larger "Operations of Nature" like limestone strata:

> The operations of nature are so simple, that when we get a glimpse of the manner in which they are effected in one instance, it is easy to extend our

> observations, in a satisfactory manner, to others of a similar nature. When we once perceive the manner in which calcareous stalactites are formed, it is easy to comprehend the way in which more regular strata of calcareous substances have been produced.[75]

Yet, such a descriptive method made it quite hard to advance prescriptive conclusions. For example, though he had personally witnessed limewater flowing into subterraneous caverns from the bowels of the earth, he offered no explanation as to how the water had been impregnated with lime. He merely stated that the matter "remains yet to be explained."[76]

Anderson's reticence to go beyond the observable facts is representative of the Baconian mindset that influenced most of his contemporaries, including Cullen, Black, and Walker. Chemistry in Edinburgh was inextricably linked to therapeutics, *materia medica*, and institutes of medicine, and this meant that there was a high evidentiary standard required for making a conclusion – otherwise there could be serious medical or legal consequences.[77] Such a method also explains why he did not follow in the footsteps of Buffon and other contemporary theorists who explained secondary and primary strata formation by postulating hundreds of thousands of years.[78] Most of the chemical reactions described in Anderson's "Essay on Quicklime" took place, at the most, within a few days (some of them took a few seconds). In the few instances where the essay makes temporal assertions, he takes care to justify his comments either by mentioning experiments that he has performed or by citing the testimony of others.[79] When speaking about the limestone of Matlock[80] and longevity of cement,[81] two topics relevant to "history" of geological strata, he cites eyewitness accounts that testify to the possible age of the cemented strata.

CONCLUSION

This essay has taken us one step closer to understanding the types of chemical principles that were relevant to georgically-minded readers in late eighteenth-century provincial Scotland. Anderson was a recognized writer on applied chemistry and his "Essay on Quicklime" was part of a multivolume series that went through at least four British editions. At this time, Scottish physicians, industrialists, and farmers employed the same chemical nomenclature that was taught in the universities. Although professors like Cullen and Black were interested in the chemistry of agriculture, most of their ideas on this subject were not committed to print and to this day the bulk of their lectures languish as manuscripts housed in the University of Edinburgh's Department of Special Collections. One of the values of Anderson's quicklime essay is that it is a published interpretation of the chemical ideas presented in their lectures (a point that also applies to the publications of other Scottish chemists like David McBride and Archibald Cochrane, ninth Earl of Dundonald). By breaking down the teachings of Cullen and Black into clear definitions and straightforward, sometimes simplistic, prose, Anderson turned their theories into practice by transferring principle-based chemistry from the laboratory into the hands of farmers and industrialists.

Yet the content of Anderson's essay also reveals that some of his readers were not chemically ignorant. His explication of lime was intended for improvement-minded farmers and landowners; and as his comments throughout his essay indicate, he realized that a number of his readers were already chemically literate. More implicitly, but just as telling, he assumed that the he did not need to defend the notion that limestone, which at face-value appeared to be solid, was actually a compound of earth and air. Since he was keen to include ideas that would make the book more marketable, this suggests that his provincial readers were amenable (or perhaps indifferent) to treating air as a separable or fixed principle. Many of them would have been familiar with the subject via the chemistry and agricultural lectures on offer in Edinburgh from the 1760s to the 1790s or via the more simplified versions of chemistry that frequently appeared in the *Scots Magazine*.[82] Establishing the larger state of provincial chemistry in Scotland has yet to be done, but Anderson's essay points to the need for studies that address how the theoretical principles of the university were modified or axiomatically extended to fit the practical needs of a local environment.

In Anderson's case, the instruments that he used to conduct *in situ* experiments were remarkably simple and the questions that guided his research were influenced by his interest in utility. Yet, as mentioned in the above sections, he was not merely an automaton who thoughtlessly manipulated the chemical ideas of his teachers. His discussion of crystallization and his appendix of chemical explanations indicate that his experiments were guided by a deeper knowledge of theoretical principles and concepts. Moreover, these thoughts were not confined to subjects that have been traditionally ascribed by historians to be the domain of late eighteenth-century chemistry. Anderson's writings remind us of an intellectual fluidity that allowed chemical concepts to migrate among medicine, agriculture, and geology. In particular, his geologically-related comments demonstrate specificity and exhibit a resourceful way of thinking that was motivated by the places and people that he visited, not necessarily by research agendas being sent down from the university. There is no doubt that he acknowledged the centrality of the chemical principles taught in the medical school, but he clearly considered himself an expert on application and this led him to formulate his own opinion on stratigraphical genesis and formation. Such a nuanced intellectual path followed the larger connotations associated with "Agricola," the pseudonym that he used to sign some of his published work. In Latin, the term means 'farmer' and it signaled Anderson's solidarity with both classical and early modern authors who considered agriculture to be part of natural history. It was this cast of mind that led him to mix the principles and practice of chemistry and which certainly earns him the right to be called the "Aberdeen Agricola."

APPENDIX

Since there were many local and personal nuances that affected the chemical classification of the minerals used by physicians and farmers, I have compiled Tables 1 and 2 so that James Anderson's work can be compared to other studies that have been written about chemistry or mineralogy of his contemporaries.

Table 1. Varieties of "Calcareous" Stones[83]

	Variety	Anderson's definition
1.	Marble	Pure calcareous matter that is perfectly crystallized
2.	Opaque marble	Pure calcareous matter that is less perfectly crystallized
3.	Chalk	Somewhat pure, but uncrystallized, calcareous matter concreted into a mass
4.	Limestone	Calcareous earth (pure and impure) that has formed crystals but which also contains sand
5.	Sparr	Pure calcareous matter that has formed into small crystals that are "of a transparent whiteness"
6.	Shell	Calcareous matter that comes from "animals" like corals and corallines
7.	Marle	Calcareous matter that has not crystallized because it is mixed with clay (and sometimes sand)
8.	Shell-marle	Whitish calcareous powder that remains after shells have lost their animal gluten
9.	Shell-sand	Small, ground-up, and gritty fragments of shells (like that found on a Highland beach)

Table 2. Earths in Edinburgh

Cullen (1750s)[84]	Black (1767)[85]	Walker (1760s)[86]
Vitrescible	Absorbent	Vitrescible
Calcareous	Gypseous	Calcareous
Argillaceous	Argillaceous	Argillaceous
(Talky) Talks	Flints	Talky

NOTES

[1] These local contexts are the focus of the essays in Bernadette Bensaude-Vincent and Ferdinando Abbri, eds., *Lavoisier in European Context: Negotiating a New Language for Chemistry* (Canton, MA: Science History Publications, 1995).

[2] Save for the works that address the Scottish years of James Watt's steam engine research.

[3] Although Roebuck established the first sulfuric acid plant in Scotland, nothing of substance has been written about his career as a chemist. Laing was a provincial minister who also held an MD. He used his chemical knowledge to publish on spas, one example being *An Account of Peterhead, its Mineral Well, Air and Neighbourhood* (London, 1793).

[4] Brief accounts of Anderson's life can be found in the *Oxford Dictionary of National Biography* (Oxford: Oxford University Press, 2004) and *The Gentleman's Magazine* 78, 1808, 1051–54.

[5] James Anderson, *Free Thoughts on the American Contest* (Edinburgh, 1776), *Enquiry into the Nature of the Corn Laws* (Edinburgh, 1777), and *Observations on Slavery* (Manchester, 1789). The larger context of Scottish agriculture at this time is addressed recently in Neil Davidson, "The Scottish Path to Capitalist Agriculture 3: The Enlightenment as the Theory and Practice of Improvement," *Journal of Agrarian Change*, 5, 2005, 1–72.

[6] James Anderson, *Observations on the Means of Exciting a Spirit of National Industry* (Edinburgh, 1777), *Miscellaneous Observations on Planting and Training Timber-Trees* (Edinburgh, 1777), *An Inquiry into the*

Causes that Have Hitherto Retarded the Advancement of Agriculture in Europe (Edinburgh, 1779), *The True Interest of Great Britain Considered: Or a Proposal for Establishing the Northern British Fisheries* (London, 1783), *An Account of the Present State of the Hebrides and Western Coasts of Scotland* (Edinburgh, 1785), *Observations on the Effects of the Coal Duty* (Edinburgh, 1792), *General View of the Agriculture and Rural Economy of the County of Aberdeen* (Edinburgh, 1794), *A Practical Treatise on Peat Moss; Considered as in its Natural State Fitted for Affording Fuel, or as Susceptible of Being Converted into Mold* (Edinburgh, 1794).

[7] *The Bee, or, Literary Weekly Intelligencer* (Edinburgh, 1790–94). Unlike Anderson's previous ventures, *Recreations* was published in London (1799–1803).

[8] The classic text for Scotland is Archibald Clow and Nan L. Clow, *The Chemical Revolution: A Contribution to Social Technology* (London: Batchworth Press, 1952).

[9] James Anderson, "On Cast Iron," read 2 August 1784, summarized in *Philosophical Transactions of Edinburgh* 1, 1788, 26–27.

[10] Samuel Parkes, *The Chemical Catechism, with Notes, Illustrations, and Experiments*, 4th Edition (London, 1810), 222.

[11] See James Anderson, "Cursory Hints and Anecdotes of the Late Doctor William Cullen of Edinburgh," *The Bee* 1, 1791, 1–10, 45–56, 121–25, and 161–66.

[12] In Scotland, these five forms seem to have been set after Macquer's *Elements de Chymie-Theorique* (Paris, 1754). Principle-based chemistry is discussed in Frederic L. Holmes, *Eighteenth-Century Chemistry as an Investigative Enterprise* (Berkeley: University of California Press, 1989), Mi Gyung Kim, *Affinity, That Elusive Dream: A Genealogy of the Chemical Revolution* (Cambridge, MA: MIT Press, 2003), and A. M. Duncan, *Laws and Order in Eighteenth-Century Chemistry* (Oxford: Clarendon Press, 1996).

[13] Joseph Black, *Notes from Doctor Black's Lectures on Chemistry 1767/8*, trans. Thomas Cochrane, ed. Douglas McKie (Wilmslow: Imperial Chemical Industries, 1966), 26.

[14] Donavan suggested that "principle" was a term in which "properties were treated as causes," Arthur Donovan, *Philosophical Chemistry in the Scottish Enlightenment* (Edinburgh: Edinburgh University Press, 1975), 114. However, Black and Cullen's use of the word shows that they also saw the word as describing a state of matter that was, nominalistically, the same as an Aristotelian "element" and that could be used for nomenclatural purposes. This second bears a notable philosophical similarity to the use of "matter states" in early quantum mechanics. See R. F. Hendry, "The Physicists, the Chemists and the Pragmatics of Explanation," *Philosophy of Science* 71, 2004, 1048–59, esp. 1049.

[15] The Aristotelian notion of "element," that is, a basic substance of matter. Henry Temple Croker, a contemporary of Black, defined "Element" by listing "Water, Air, Oil [inflammable], Salt, Earth" in *The Complete Dictionary of Arts and Sciences: In which the Whole Circle of Human Learning is Explained,* 3 vols (London, 1764–66), s.v.

[16] In the early modern period, "substance"(Latin *substantia*) generally signified "a separate or distinct thing"; *Oxford English Dictionary*, def. 2. Though Joseph Black used the words substance and body interchangeably, his lectures infer that he thought that a substance was a measureable piece of matter that existed in two forms: solid (which he often used for bodies) and fluid. Black, *Lectures 1767/8*, 3 and 8.

[17] Black, however, recalibrated parts of the table. See Donovan, *Philosophical Chemistry*, 218.

[18] Robert P. Multhauf, *The Origins of Chemistry* (Reading: Gordon and Breach, 1993), 344.

[19] M. D. Eddy, "Set in Stone: The Medical Language of Mineralogy in Scotland," in David Knight and M. D. Eddy, eds., *Science and Beliefs: From Natural Philosophy to Natural Science* (Aldershot: Ashgate, 2005), 77–94.

[20] A. H. Maehle, *Drugs on Trial: Experimental Pharmacology and Therapeutic Innovation in the Eighteenth Century* (Amsterdam: Rodopi, 1999).

[21] Joseph Black, *De Humore Acido a Cibis orto, et Magnesia Alba* (Edinburgh, 1754).

[22] Joseph Black, "Experiments upon Magnesia Alba, Quick-lime, and Other Alcaline Substances," *Essays and Observations, Physical and Literary* 2, 1756, 157–255.

[23] Donovan, *Philosophical Chemistry*, 201–21.

[24] Fred Page, "Lime in the Early Bleaching Industry of Britain, 1633–1828: Its Prohibition and Repeal," *Annals of Science* 60, 2003, 185–200, esp. 187.

[25] David MacBride, *Experimental Essays on Medical and Philosophical Subjects* (London, 1767).
[26] Francis Home, *Experiments on Bleaching* (Dublin, 1771). Black's essay was entitled "An explanation of the effect of lime upon alkaline salts, and a method pointed out whereby it may be used with safety and advantage in bleaching." Black spoke favorably of MacBride's work in his *Lectures on the Elements of Chemistry*, 2 vols (Edinburgh, 1803), 2, 97.
[27] The larger context in which science was fused with Scotland's nationalistic brand of utilitarianism is addressed in C. W. J. Withers, *Geography, Science and National Identity: Scotland Since 1520* (Cambridge: Cambridge University Press, 2001). See also Jan Golinski, *Science as Public Culture: Chemistry and Enlightenment in Britain, 1760–1820* (Cambridge: Cambridge University Press, 1992), esp. chapter 2.
[28] James Anderson, *Essays Relating to Agriculture and Rural Affairs*, 4th ed., 3 vols (London, 1797–98), 1, 408. For the remainder of this essay, I will use this edition of the 'Essay on Quicklime'. It was also published in earlier editions of the work, but the 4th edition contains several notable footnotes on the new nomenclature that I will address is later sections. The main text remained unchanged in all editions.
[29] Arthur Donovan, "Scottish Responses to the New Chemistry of Lavoisier," *Studies in Eighteenth-Century Culture* 9, 1979, 237–49 and "The New Nomenclature among the Scots: Addressing Novel Chemical Claims in a Culture under Strain," in Bensaude-Vincent and Abbri, *Lavoisier in European Context*, 113–21.
[30] Anderson, *Agriculture*, 393.
[31] See Anderson's chemical definitions in the appendix.
[32] There were unsuccessful efforts both in Scotland and England during the 1750s that sought to propose a unit of common weight and this resulted in more disagreement until the introduction of the Imperial system in the 1820s. A. D. C. Simpson and R. D. Connor, "The Mass of the English Troy Pound in the Eighteenth Century," *Annals of Science* 61, 2004, 321–49; R. D. Connor and A. D. C. Simpson, *Weights and Measures in Scotland: a European Perspective* (Edinburgh: National Museum of Scotland, 2004).
[33] Anderson, *Agriculture*, 397. Here Anderson uses the word "lime" to represent "quicklime."
[34] Black also did this at times in his lectures (or perhaps the students taking the notes conflated the terms); Black, *Lectures 1767/8*, 61.
[35] Anderson, *Agriculture*, 409.
[36] Ibid., 493.
[37] See Table 1 in the Appendix.
[38] Anderson, *Agriculture*, 399.
[39] Donald Monro, who lived in London, lists Cullen's table of salt affinities in "An Account of Some Neutral Salts Made with Vegetable Acids, and with the Salt of Amber," *Philosophical Transactions* 57, 1767, 479–516.
[40] Donovan, *Philosophical Chemistry*, 122.
[41] A "Salt" at this time included acids and alkalis and Cullen's experiments were linked into a much larger European research tradition interested in these substances. In Cullen's case, he was specifically building on the works of Joan Baptista Van Helmont, John Mayow, Georg Ernst Stahl, Charles DuFay, Paul Malouin, Pierre Macquer, Duhamel du Monceau, Frederick Hoffmann, and William Brownrigg; see Donovan, *Philosophical Chemistry*, 122–26. For the work of William Cadogan (1711–97) and Hoffmann on this topic, see A. B. Davis and J. B. Eklund, "Magnesia Alba before Black," *Pharmacy in History* 14, 1972, 139–46.
[42] Cullen could not decide whether gypsum was a salt, a dilemma repeated by Anderson, who called gypsum is an "earthy salt," Anderson, *Agriculture*, 449. Cullen's student, John Walker, also held that there were "earth acids." For both Cullen's and Walker's thoughts on this matter, see M. D. Eddy "Scottish Chemistry, Classification and the Late Mineralogical Career of the "Ingenious" Professor John Walker (1779–1803)," *British Journal for the History of Science* 37, 2004, 373–99, esp. 416–17.
[43] M. D. Eddy, "The Doctrine of Salts and Rev. John Walker's Analysis of a Scottish Spa," *Ambix* 48, 2001, 137–60.
[44] Table 2 in the Appendix gives a chart that contains the different types of earths accepted by Cullen, Black, and Walker. As Anderson stated: "Absorbent earths are all those that unite with acids, of which there are several varieties; calcareous earths being one of these;" Anderson, *Agriculture*, 575.
[45] This was most probably because such an approach was more pedagogically expedient, though it also could be that he remained faithful to Cullen's classification.

[46] Black's *Elements,* used "Absorbent" and "Alkaline Earth" synonymously (see 2:23). It is not clear whether this was a recalibration added by the book's editor, John Robison. Student notes tend to use the term "Absorbent"; however, these notebooks were often compilations and it is sometimes hard to tell exactly when they were written down. Black usually lectured on absorbents in lectures (*circa*) 60 to 70. For comparison, see Henry Beaufy (transcriber), *Manuscript Copy of Lectures in Chemistry Given by Joseph Black, Professor of Medicine and Chemistry, Edinburgh University, 1766–1799 [c. 1771–1775], Volume IV*, Aberdeen University Library Special Collections MS 38185.

[47] John Walker, *Schediasma fossilium* (Edinburgh, 1781), *Classes fossilium* (Edinburgh, 1787), and *Systema fossilium* (c. 1797), Glasgow University Library Special Collections Department, Gen. 1061.

[48] Anderson, *Agriculture*, 451.

[49] This analytical method is treated in more detail in William R. Newman and Lawrence M. Principe, *Alchemy Tried in the Fire: Starkey, Boyle, and the Fate of Helmontian Chymistry* (Chicago: University of Chicago Press, 2002), esp. chapter 2; W. R. Albury and D. R. Oldroyd, "From Renaissance Mineral Studies to Historical Geology, in the Light of Michel Foucault's *The Order of Things*," *British Journal for the History of Science* 10, 1977, 187–215.

[50] Anderson, *Agriculture*, 507.

[51] Though acids from this time were not as pure as they are today, the modern equivalents of those above are *aquafortis* = nitric acid (HNO_3) and spirit of salt = hydrochloric acid (HCl). Equivalency tables that convert eighteenth-century substances to modern nomenclature can be found in Jon Eklund, *The Incompleat Chymist: Being an Essay on the Eighteenth-Century Chemist in his Laboratory, with a Dictionary of Obsolete Chemical Terms of the Period* (Washington: Smithsonian Institute Press, 1975); Torbern Bergman, *A Dissertation on Elective Attractions*, A. M. Duncan, ed. (London: F. Cass, 1970), appendix 2. The appendix at the end of Volume 2 of Black's *Elements* gives conversion tables from Black's terms to the new nomenclature as it stood ca. 1800.

[52] Anderson, *Agriculture*, 508.

[53] The production of acids during the eighteenth-century is treated briefly in K. Warren, *Chemical Foundations: The Alkali Industry in Britain to 1926* (Oxford: Clarendon Press, 1980).

[54] Anderson, *Agriculture*, 553.

[55] Ibid., 512.

[56] Ibid., 536.

[57] For industry and agriculture, see Clow and Clow *The Chemical Revolution*; for pharmacology, see D. L. Cowan, *Pharmacopoeias in Britain and America, 1618–1847* (Aldershot: Ashgate, 2001) and J. R. R. Christie, "William Cullen and the Practice of Chemistry," in A. Doig, J. P. S. Ferguson, I. A. Milne and R. Passmore, eds., *William Cullen and the Eighteenth Century Medical World* (Edinburgh: Edinburgh University Press, 1993), 98–109.

[58] Glasgow University Library, MSS Box 7.3.

[59] William Cullen. *Abstract from Dr. Cullen's Lectures on Agriculture*, John Walker (transcriber) (c. 1766), Edinburgh University Library Special Collections, Dc.3.70; *Collection of Notes and Extracts Relating to Farming and Husbandry: With Lectures on Vegetation and Agriculture*, James Cunningham (transcriber) University of Aberdeen Special Collections, King's College MS 564/2; it is highly likely that Anderson also sat in on these lectures.

[60] William Cullen, *The Substance of Nine Lectures on Vegetation and Agriculture, Delivered to a Private Audience in the Year 1768 ... With a few notes by George Pearson* (Edinburgh, 1796). The contents and larger relevance of Cullen's agriculture lectures is addressed by C. W. J. Withers, "William Cullen's Agricultural Lectures and Writings and the Development of Agricultural Science in Eighteenth Century Scotland," *Agricultural History Review* 37, 1989, 144–56.

[61] Black commented on "good" and "bad" marl as early as the 1760s; Black, *Lectures 1767/8*, 58. Agricultural topics were also covered by leading chemists like Herman Boerhaave and Pierre Macquer, both of whom were required reading for chemistry course run by Cullen and Black. Boerhaave's interest in the chemical composition of plants is addressed in Ursula Klein, "Experimental history and Herman Boerhaave's chemistry of plants," *Studies in the History and Philosophy of the Biomedical Sciences* 34, 2003, 533–67.

[62] Black, *Elements*, 2:93–94.

[63] The first several of Walker's mineralogy lectures covered soils. For his attempts to secure the Chair of Agriculture, see C. W. J. Withers, "A neglected Scottish agriculturalist: The "Georgical Lectures" and Agricultural Writings of the Rev. Dr. John Walker (1731–1803)," *Agricultural History Review* 33, 1985, 132–43.
[64] A point mentioned throughout Roy Porter, *The Making of Geology* (Cambridge: Cambridge University Press, 1977).
[65] For contemporary accounts of stalactites and caves, see Black's comments on stalactites in his *Lectures 1767/8*, 57, and in *Elements*, 2, 24–28; Walker also addressed caves in his natural history lectures, *Lectures on Geology, Including Hydrology, Mineralogy, and Meteorology with an Introduction to Biology*, ed. Harold W. Scott (Chicago: University of Chicago Press, 1966), 188.
[66] Anderson, *Agriculture*, 414.
[67] Ibid., 417.
[68] Ibid., 407. He also notes that crystals form more easily when the saline liquid is heated, which could possibly be linked to Cullen's thoughts on the material basis for heat and coldness. See William Cullen, "Of the Cold Produced by Evaporating Fluids, and of Some Other Means of Producing Cold," *Essays and Observations, Physical and Literary* 2, 1756, 145–56.
[69] Anderson, *Agriculture*, 422.
[70] See Eddy, "Professor John Walker."
[71] These were included in *Essays Relating to Agriculture*; vol. 1 (1797) contains extended essays entitled "Of Enclosures of Fences," "On Draining Bogs and Swampy Grounds," and "On the Proper Method of Levelling High Ridges." The practices outlined in these works could potentially increase the productivity of land several times over. To this end, Anderson was keen to protect the originality of his innovations (chemical or otherwise). See the anonymous review of the revised 1797 version of his bogs essay in the *Scots Magazine* 60, 1798, 840–41.
[72] Walker's mineralogy and geology lectures suggest that he held there were a series of floods that happened at different times and created different saline solutions (and hence, different types of cementation). In his introductory comments to his lectures on earths, Black stated that the "materials" of geological strata had been "arranged by water, depositing or arranging them one over another, in succession." Black, *Elements*, 2, 11–12.
[73] By the late eighteenth century, in Scotland at least, this flood was differentiated from the Biblical flood involving Noah. The chemical underpinnings of this predominant rationality are discussed in Rachel Laudan, *From Mineralogy to Geology* (Chicago: Chicago University Press, 1987); G. L. Herries Davies, *The Earth in Decay: A History of British Geomorphology, 1578–1878* (London: Macdonald and Co., 1969); M. D. Eddy, "Geology, Mineralogy and Time in John Walker's University of Edinburgh Natural History Lectures," *History of Science* 39, 2001, 95–119.
[74] Black called chemistry an "art" for his entire career.
[75] Anderson, *Agriculture*, 416.
[76] In later editions of the text he did note that "Chemical Philosophers" offered pneumatic explanations for this phenomenon, but he then stated that it was unnecessary to recount their theories because he felt that it was irrelevant to the focus of his argument; Anderson, *Agriculture*, 413.
[77] Andrew Duncan, senior (1744–1828) lectured on medical jurisprudence at the University of Edinburgh during the 1790s, and in 1796 the medical faculty created a professorship for the subject.
[78] The epitome of this view was expressed in Georges Louis Leclerc Buffon, *Natural History, General and Particular*, William Smellie (trans.) (London: Strahan and Cadell, 1781). Buffon extended the earth's history beyond several thousand years and his theoretical approach had a particularly bad reputation in Edinburgh. See Black, *Elements*, 2,15–17.
[79] Anderson explicitly states that he had performed several of experiments on his own. For an example of his own experiments on slaked lime, see Anderson, *Agriculture*, 410. Crystal formation was witnessed by himself (410) in his own experiments and by professors in the medical school.
[80] He avers that "within the memory of man," travellers in Matlock had witnessed the "rapid" growth of calcareous strata; ibid., 417.
[81] On this point he notes the "lime-cement employed by the Ancients" in Scotland "appears to be much firmer than that which has been made in modern times," ibid., 435.

[82] Although no formal study has been published on the chemical content of the *Scots Magazine*, a cursory survey of the volumes published between 1750 and 1800 shows that applied chemistry was discussed in almost every issue. Some of these articles mention chemistry only in passing, while others were meant to educate. Even in 1800, lime was still a popular topic; see for example, "A Durable Cement," *Scots Magazine* 62, 1800, 176; "Lime as a Manure," ibid., 399; "[A Patent] for Preparing the Oxygenated Muriates of Limes," ibid., 424; "[New Patent] for a Cement," ibid., 714.

[83] Anderson, *Agriculture*, 400–03.

[84] William Cullen, "Of Vitrescent Earths and Vitrifications... by Cullen," Glasgow University Library MS Cullen 268/8; "A Chemical Examination of Common Simple Stones & Earths ... by William Cullen with Notes [Incomplete] on Alkali Earths and the Earth's Structure," Glasgow University Library MS Cullen 264, fol. 1.

[85] Black, *Lectures 1767/8*, 56–67.

[86] John Walker, *Adversaria* (1766–72), Bound MS, Glasgow University Library MS Murray 27, f. 157.

TREVOR H. LEVERE

DR. THOMAS BEDDOES (1760–1808): CHEMISTRY, MEDICINE, AND BOOKS IN THE FRENCH AND CHEMICAL REVOLUTIONS

Dr. Thomas Beddoes was a mere bit-player in Britain's cultural history, as Roy Porter has told us;[1] but his career, seen through correspondence, notes, publications, and instruments, invites us to reassess several key aspects of the history of late eighteenth-century chemistry. He was not highly skilled in the chemical laboratory. His main medical research project, based on the hope that different gases would have therapeutic effects on a variety of disease, was a failure. Nonetheless he continues to tantalize historians of chemistry and medicine. There are already three biographies of him, and a further study is in preparation.[2] Here, I wish to examine his medical chemistry or chemical medicine, his role as a truly European chemist and physician, his engagement with industrial chemists and engineers in the Lunar Society of Birmingham, and the chemical, medical, and political networks to which he belonged or contributed. Iatrochemistry has been little considered in the late eighteenth century, but it was important, in ways that Beddoes helps us to understand. Nationalism is often identified with the different parties engaged in the Chemical Revolution, but the case of Thomas Beddoes makes it clear that internationalism is also important. The debates between the new and the old chemistry are often made to revolve around two issues, phlogiston and quantification; Beddoes helps us to understand that the chemical debates sometimes had a much broader and more complex underpinning.

EUROPEAN MEDICINE AND CHEMISTRY: BOOKS AS AMBASSADORS

The son of a tanner, Beddoes taught himself to read Dutch, French, German, Italian, Latin, Portuguese, Spanish, and probably Swedish; he completed his acquisition of foreign languages as an undergraduate student at Oxford from 1774 to 1779. In those years, he published his translation of an Italian work on animal and vegetable physics by Lazzaro Spallanzani.[3] He also translated from the Latin Bergman's treatise on elective affinities, and Scheele's chemical works.[4] From Oxford, he removed to Edinburgh to study medicine; the medical curriculum there followed Boerhaave's earlier model in giving a prominent place to chemistry. Edinburgh in the late eighteenth century was a leading European university in both fields, unlike the more introverted English universities; Joseph Black's students came from all Europe and beyond,[5] including Russia, whither Catherine the Great tried to entice him. Black was not fond of travel, and declined to

L. M. Principe (ed.), New Narratives in Eighteenth-Century Chemistry: Contributions from the First Francis Bacon Workshop, 21–23 April 2005, 157–176.

go.[6] His foreign students brought with them lecture notes from home, and Beddoes was soon borrowing manuscript lectures of "the best Professors on the continent," and taking notes from them.[7] In 1786, he returned to Oxford to obtain his MD, and in the following year travelled on the continent. He was with Guyton de Morveau in Dijon when Lavoisier and Fourcroy arrived with their wives, along with Monge and Vandermonde.[8]

In 1787, Beddoes returned to give chemical lectures in Oxford. He wrote to Black,[9] lamenting the lack of a good elementary textbook to use with his own lectures. He told Black that he had collected all the modern elementary books, including the German ones, "which are not a few." In 1800, he wrote to a correspondent that he was "engaged in the composition of a work on popular physiology and preventive medicine, in which I would derive all the aid I can from the whole literature of Europe."[10] In chemistry and medicine, this was truly a lifelong European purview.

He did not, however, limit his interest in books to chemistry and medicine, as the sale catalogue of his library makes plain. In November 1809, almost one year after his death on 23 December 1808, an auction was announced in *A Catalogue of the very valuable and extensive library of Thomas Beddoes, M.D. of Clifton, near Bristol, lately deceased: containing a very capital collection of modern publications in all the departments of surgery and medicine, voyages and travels, antiquities, natural history, and belles lettres: likewise all the late best German writers on the above subjects, which will be sold by auction by Leigh and S. Sotheby, Booksellers, at their House, No. 145, Strand, opposite Catherine Street, On Friday, November 10, 1809, and Nine following Days, (Sundays excepted) at 12 o'Clock*. The auctioneer's copy of the catalogue, identifying purchasers for each lot, and giving prices, was in the British Library.[11] It was a remarkable library for a West-country Doctor. The only major field of Beddoes's interests not prominent in the catalogue is that of politics. There were 2131 lots. The first five days saw the sale of 1071 of them, all but one in German (the odd one out was in Dutch), mainly devoted to chemistry and medicine, but with a very considerable number of works of literature, history, and philosophy. It was Beddoes who encouraged the young Samuel Taylor Coleridge to go to study in Germany.

Beddoes once had a visitor, a Dr. Frank from Vienna. Frank was enjoying the freedom to travel in the brief window of peace that followed the Treaty of Amiens in 1802. He met Richard Lovell Edgeworth, Beddoes's father-in-law, and, armed with an introduction from Edgeworth, arrived in Clifton, the suburb of Bristol where Beddoes lived and had his pneumatic practice.[12] There were at that time at least three Dr. Franks from Vienna who had published one or more medical books. When Beddoes greeted his visitor, he did so from behind an armful of books, arranging them in piles, each by a different Dr. Frank, and asked "Which Dr. Frank are you?" The sale catalogue lists 26 volumes by Austrian and German Dr. Franks. It lists many hundreds of volumes dealing with differrent areas of medicine, including pneumatic medicine (there is a German translation of one of Beddoes's own books on consumption and its treatment).[13] There are also monographs and textbooks by Scheele, Ritter, Volta, Gren, Klaproth, and many others, and a run of Crell's *Chemische Annalen* starting in 1784. The French books in the sale cover the standard up-to-date range, including Lavoisier, Berthollet, Chaptal, Fourcroy, Guyton, and Kirwan (in the French translation made by Mme. Lavoisier, with detailed point-by-point rebuttals by Lavoisier and others).[14] The remaining five days saw the sale of books in English, French, German, Greek, Portuguese, Latin, Spanish, Italian, and Dutch. The English books are also comprehensive in chemistry, and what one would expect, save that Priestley is represented only by his *Disquisitions on Matter and Spirit*.[15]

It is worth pausing for a moment to consider the circumstances in which Beddoes built his library. Most of Beddoes's professional career took place in wartime, from the American Revolution (or War of Independence), which broke out while he was an undergraduate and took on the dimensions of a world war, to the Napoleonic Wars, which continued beyond his death. Some civilities were maintained during these conflicts. Joseph Banks negotiated the liberation or exchange of British and French scientists.[16] Interested members of Parliament managed to get hold of French periodicals and political reports. The regular international mail, however, largely dried up. Trading vessels had to negotiate the blockade of Napoleonic France. In an admittedly extreme case, it took six years to ship one set of philosophical and chemical apparatus from the Low Countries to the new university in Russian-controlled Dorpat (now Tartu, Estonia).[17]

In spite of these difficulties, Beddoes managed to acquire foreign books. Lavoisier sent him a copy of the antiphlogistonists' response to Kirwan in 1788.[18] Beddoes continued steadily to acquire French books published up to 1792, but in the rest of that decade only a few. Unsurprisingly, not much was published by French chemists in the years of the Terror and its immediate aftermath. Fourcroy, who managed effectively to distance himself from Lavoiser, and who alone among prominent French chemists managed to write and publish steadily through the 1790s, seemingly unfazed by Revolution and Terror, was an exception.[19] War or no war, however, Beddoes managed to keep a stream of German books coming. In this enterprise, he was helped greatly by Matthew Robinson Boulton, son of Matthew Boulton, James Watt's partner in the business of building pumps, steam engines, and buttons for army uniforms. Boulton kept a German branch office going throughout the wars. Beddoes arranged for a discretionary commission to Martini, presumably a successor to the Leipzig bookseller Johann Christian Martini. He used Boulton's office in Hamburg to pay Martini ready money, and to receive and forward the books to London or Birmingham, and thence to Clifton.[20] Beddoes did not need to own books, but he did need to read them, and in addition to creating his own library, he was a frequent borrower of books from other people. He bothered booksellers, including the French importer of foreign books Joseph DeBoffe[21] in Soho, London, and members of parliament who had access to diplomatic bags from Europe and free postage within Britain. He mercilessly made use of his close friend and former student Davies Giddy, sometime Sherrif of Cornwall, and longtime MP. After Beddoes's death and his own marriage, Giddy changed his name to Gilbert, and went on to become Humphry Davy's successor as President of the Royal Society of London. Giddy frequently facilitated Beddoes's purchases and borrowing of books. And of course, Beddoes used lending libraries, although in 1794 the Bristol Library Committee "doubted whether he was a fit person to be entrusted with books."[22] We do not know whether the Committee changed its mind on that score, but the Bristol Library records tell us of Beddoes's borrowings from 1798 to 1803, including issues of the *Philosophical Transactions of the Royal Society of London*, which he had to borrow since Banks's hostility guaranteed that he would never become FRS, even though he published four papers in the *Philosophical Transactions*. He borrowed a wide range of medical and scientific works.[23]

One other source of books for Beddoes was reviewing. Between 1793 and 1801 he wrote reviews of over 160 titles for the *Monthly Review*.[24] Sixteen of the titles he reviewed are French, among them 11 on chemical subjects; 28 are German, 10 of those books are on chemistry; 2 are Italian. Most of the French reviews are of journals; the only French chemist to appear in these reviews is Fourcroy.[25] There was a significant death-roll among French scientists in those years.[26]

Given his own success in obtaining the books he needed, he had little sympathy with librarians who failed to perform their jobs properly. In 1787, back in Oxford from his European visit, he addressed a swingeing list of criticisms to the Reverend John Price, Bodley's Librarian. The criticisms filled a 98-page booklet, entitled: *A memorial concerning the state of the Bodleian library, and the conduct of the principal librarian. Addressed to the curators of that library, by the chemical reader*. Beddoes complained, accurately, that serials were not maintained, sets were incomplete, and poor English translations were bought instead of the French originals. There were few acquisitions of German books. No care was taken "to supply us with the authors of a country, who may justly contest the palm of science and literature with those of any other nation. It may be said, indeed, that if we consider the small number of readers of German, that the use of books written in that language would be very limited. But in a place of education, I think it is rather to be expected, that the means of making literary acquisitions should be provided, than that persons should come already furnished with them."[27] Beddoes had a slew of criticisms about the Librarian, and he ended by comparing the Bodleian Library unfavorably with the libraries of the universities of Edinburgh and Göttingen. "I can discover no reason why an English should be inferior to a Scotch or an Hanoverian University, in any respect; nor why the nation should be without as ample a repository of all kinds of knowledge ..."[28] Beddoes would have been delighted to know that Göttingen University has 54 of his publications in its library today, several in German translation. Oxford has 60; Edinburgh has only 28.

Books were important to Beddoes, except for novels, which he found unsettling and which he regarded as positively unhealthy, especially for women.[29] His library and the books he borrowed were his pass-key to the international world of learning. He did not allow revolutions or wars or delinquent librarians to block his access to them.

CONVERSION TO THE NEW FRENCH CHEMISTRY

While Beddoes was a student at Edinburgh, he wrote to his friend and near contemporary, the surgeon Charles Brandon Trye, giving him thumbnail sketches of the chemistry lectures. Black's course impressed him greatly; it was "the best I have ever heard or ever shall hear. But as you cannot see his exps. in a MS, you may gain all the information this would give you from Bergman and the new edition of Macquer's dictionary when it is published. I ought perhaps to except the doctrine of latent heat, but of this I have a copy ..."[30] Black refused to publish his lectures during his lifetime; he died in 1799. Only then could his student John Robison produce an edition based upon manuscript notes.[31] But sets of notes taken by students circulated in Black's lifetime, and several of those sets have survived.[32] Beddoes regarded Black as by far the best chemistry lecturer in Edinburgh, but he was also impressed by James Gregory's chemical lectures.[33] "I shall have some notes of them," he told Trye.

Having left Edinburgh, obtained his MD from Oxford, and spent some months in France, Beddoes returned to England, convinced by his discussions with Guyton and Lavoisier that the old theories of chemistry were wrong, and that the system embryonically embodied in the new French nomenclature was right. Back in England, Beddoes began to have some doubts about the new French system. He moved to Oxford in November 1787.[34] In preparing his lectures, he had three immediate needs: a useful introductory textbook, a set of demonstration apparatus, and laboratory skills, either his own or those of an assistant or demonstrator. As he explained in a letter to Black,[35] chemistry had reached the point "when the necessity of a new arrangement appears evident without being possessed of sufficient information to discern clearly what that arrangement ought to be." Black had made very much the same point in his own lectures. He explained that he did not use a textbook for his course, because "I do not find any in which the arrangement is the best and the views are suficiently extensive and therefore I use an arrangment of my own."[36] The chemistry of gases, arising from a sequence of discoveries and theories in which Black's work on fixed air was seminal, made it hard to maintain the truth of the phlogiston theory. Beddoes told his mentor that subsequent discoveries in pneumatic chemistry seemed to suggest better hopes, "except to those who have had the folly or the misfortune to fix their opinions inalterably." That was never Beddoes's misfortune. Humphry Davy, Beddoes's new protégé in 1798, wrote admiringly "I consider Dr. Beddoes as the most truly liberal candid and philosophic physician of the age, as he has sufficient [?] to give up his own theories, as well as those he has adopted whenever they appear contrary to facts."[37] Beddoes explained to Black that the discoveries of Cavendish, Lavoisier, and others "lead to three important changes, (1) to the rejection of phlogiston, (2) and consequently no longer to consider inflammable bodies as containing one common principle (which ... leads me to <*reject*> ... the class), and (3) to place the elastic fluids or the greatest part of them before the acids immediately after the doctrine of heat. I would not make a class of elastic fluids any more than of inflammables. That state, as you have taught us, depends upon the combination of fire and the qualities of the body which was before solid or liquid may be (very) different in other respects." Here, however tentatively, was a basis for revising and developing a new arrangement and method of chemistry. Beddoes did not expect Black to accept his proposals. Black was, after all, cautious enough to avoid using the new French nomenclature, not because he knew that it was wrong, but because he did not know that it was wholly right. His lectures were in this respect conservative. Beddoes, although impressed by the new nomenclature, was also far from persuaded that it was entirely right. *Calorique*, the matter of heat, was a difficulty. Azote was unsatisfactory as a name for what we now call nitrogen, because there were many other gases that would not support life. But in spite of these difficulties, the French chemists were on the right track. "Mr. Cavendish has abandoned phlogiston and I think we are all here about to turn our backs on poor Stahl."

In the event, Beddoes did not publish his new arrangement of bodies until 1798, when he used it for his chemical lectures in Bristol; it was reprinted and more widely distributed in 1799.[38] Early work on photosynthesis and photochemistry, along with the latest work by Humphry Davy,[39] found a privileged place for light in class I of bodies. Beddoes was able to take advantage of the recent researches of Galvani and of Volta, the latter widely discussed well before they were printed in the *Philosophical Transactions of the Royal Society of London* in 1800;[40] the electric and galvanic fluids were also included. Class II contained oxygen and its suppositious combination with light, i.e. "phosoxygen." class III included hydrogen, which combined with oxygen to form water, and "azote, forming with phosoxygen nitrous air." Clearly Beddoes was taking much from the new French chemistry, but not wholesale. He omitted caloric because of Davy's researches on heat and light, and because of Lavoisier's "strange abuse" of the doctrine of latent heat. As we saw earlier, Beddoes had notes of Black's lectures on latent heat, and on this subject agreed with his master. Black was skeptical about any material theory of heat (Lavoisier's caloric embodied such a theory). Beddoes, like Black, considered that the role of heat in chemistry and in changes of state was "much more easily reconcilable to the *mechanical* than the *chemical* doctrine of heat."[41] But he was less cautious than Black in dismissing the material theory of heat, and indeed in almost every other respect. Black had noted that "many have been the speculations and views of ingenious men about this union of bodies with heat. But, as they are all hypothetical, and as the hypothesis is of the most complicated nature, being in fact a hypothetical application of another hypothesis, I cannot hope for much useful information by attending to it."[42]

When Beddoes came to the metals, he engaged in some general reflections about the nature of simple substances:

> Concerning the composition of metallic substances, not much can be said. Whether to create a diversified system of bodies out of one, or out of a few or many elements, imply most wisdom and power, is a question which different persons would decide according to their various taste in world-making. But we have some indications that the metals are not so many simple substances; and in the case of some among them, by accurate *close* experiments on organized bodies, we might have certain proofs. The existence of iron in such variety of plants and animals; and of manganese in some plants, suggests an opinion that these metals are compounded by the organic powers; and then we are warranted by analogy in surmising that the other metals consist but of the same principles differently modified.

He conjectured further that soda and potash were probably complex; even Lavoisier had allowed as much. Beddoes concluded with the reflection: "Should the present view lead, in but a single instance, to successful experiments on the decomposition of bodies of unascertained constitution, it would be a great advance towards the removal of the present difficulties in chemical theory and practice."

Beddoes learned from Davy, and Davy learned from Beddoes; it is very hard not to regard some of Davy's triumphant electrolytic decompositions in the 1800s, most notably those of the fixed alkalis, as prompted by his work with Beddoes. The same is true of his later speculations about the compound nature of the metals and the ultimate simplicity of matter.[43]

Back in 1787 Beddoes was uneasy about the new French system and nomenclature, but he was far from ready to create his own system in their place. Nevertheless, in 1788 he still planned to write an elementary textbook, and sought Black's permission to dedicate the book to him. "I shall always consider you are the person to whom I not only owe any just views I may entertain of chemistry, but of every other subject. ... It is for these reasons that I wish your name to adorn any sketch of a system of chemistry which I may offer to the public. For as I am in no haste to be rich, I choose rather to pay the debts of gratitude than to carry offerings to the vanity of the great."[44] Beddoes wrote nearly 70 books and pamphlets, but he never produced an introductory textbook of chemistry, perhaps because he could never settle in his own mind the *details* of any new system. He knew that the French were on the right track, but also believed that their system was flawed. He had a high regard for Lavoisier and Guyton, and equally for Cavendish and Priestley. The differences among them on grounds of theory did not worry Beddoes; theories were there to be useful, and to aid discovery – they were never fixed and final. But experiment, and above all interpreting the results of experiment, was something else. He told Black: "Of the chemical news which I have lately heard, the most important is, that Dr. Priestley is repeating Mr. Cavendish's exps on water with airs as dry as possible and instead of water, he gets an acid .... Dr. Withering has the acid to examine, but has not found out what it is. It is not the nitrous, or seems not to be it."[45] One year later, he was beginning to be persuaded that the French chemists were wrong: "Dr. Priestley seems totally to have overthrown the antiphlogistic theory – I am anxious to hear what the French chemists have to say on the other side – I have seen some of their private objections to Dr. Priestley's inferences, but they are totally insignificant – still however we owe much to Mr. Lavoisier for having taught us the combinations of pure air – I now suspect that both you and he must have overlooked some acid in burning spirit of wine. It certainly cannot be pure water that is formed."[46] How to turn data into evidence is a perennial problem in science, and one which led Beddoes into a flurry of reckless theorizing and precipitate retractions.

But here he was, lecturing in Oxford to an audience that, while smaller than Black's in Edinburgh, was, according to Beddoes, larger than any "seen at Oxford, at least within the memory of man, in any Department of knowledge."[47] Writing a textbook that would supersede all previous ones was only one of his challenges. Some chemists seem to have the chemical equivalent of a gardener's green thumb, which makes everything prosper. Henry Cavendish was one such chemist. Beddoes, however, was neither an experienced nor a highly skilled experimentalist. He appealed to Black:

> What I find most difficult is to repeat some of those apparently simple exps. which in your hands are so striking and so instructive. I have not yet learned how to show the gradual approach towards saturation by throwing slowly a powdered salt into water. What salt do you use? & how do you perform the expt? How do you contrive to make that capital expt which shews the burning of iron in dephd air? I mean to attempt it, but am told that the vessel has been frequently in other hands burst with great violence? Do you put sand at the bottom? I know the form of the vessel &c. What salt do you use to shew the effects of agitation upon mixture?[48]

Looking back on his Oxford lectures in the summer after leaving Oxford, Beddoes despondently admitted to Thomas Wedgwood, his friend and patient, son of the potter Josiah and one of the first to explore photochemistry, that "I know very well that some of my chemical lectures at Oxford were dull – The subject is often so to those who look only to be entertained by showy experiments."[49] A second-hand report of Beddoes's lectures, by one who had been a student when Beddoes was Reader, but who never attended them, nonetheless fits Beddoes's depressed account. "He was a man of science and of genius, but by no means successful as a lecturer. His figure and delivery were ungraceful, his language inflated and ambitious, and he was so singularly awkward in the mechanical part of his experiments that they generally failed, and he was then compelled to proceed in his discourse on the hypothesis that the result had been the reverse of that which the eyes of his audience would have led them to believe."[50]

A theoretical framework was difficult to establish while contradictory reports of experiments were circulating; there was no textbook that fitted Beddoes's approach to chemistry, which grasped at too many different analogies to be coherent; and he needed to build up a collection of demonstration apparatus. By April 1791 he felt that at any rate he had a suitable collection of apparatus. The completion of this collection went along with the end of his vacillation between Priestley and Lavoisier, between dephlogisticated air and oxygen. Indeed, acquiring and building the apparatus that he needed in order to repeat Lavoisier's experiments was in itself a major step towards reaching a decision about the rival theories. Apparatus is designed to function within a theoretical framework, and the results obtained with that apparatus tend accordingly to validate the theory; there is a kind of circularity tying theory to apparatus via experiment, and sometimes the demonstration of new theories depends upon the use of new apparatus. That was what Lavoisier himself believed when he wrote his *Traité* of 1789, giving a third of the book to the description of apparatus and instructions for using them: "perhaps, if, in the different papers that I have given to the Academy, I had dwelt at greater length to the details of manipulation, I would have made myself more easily understood, and the science would have made more rapid progress. .... It will easily be perceived that this third part could not have been extracted from any existing work, and that, in its principal articles, I could only be helped by my own experience."[51]

In 1789, Beddoes had acquired a burning glass from Parker, the leading supplier of scientific glassware.[52] That was an instrument equally useful for experiments in Priestley's phlogistic scheme and in Lavoisier's antiphlogistic one. But in 1791, a year in which he ordered chemical apparatus of the same size and composition as "Dr. Priestley's last tubes and retorts,"[53] he told Black that he had "a very valuable assortment of chemical apparatus – a gazometer very much improved upon Mr. Lavoisier's[54] &c so that I am able to show any and every expt. in his book – It has been constructed by a pastry cook in this place, a perfect prodigy in mechanics, who has invented and executed an improved barometer of which the column of [mercury] is not altered by temperature; and by the help of which I can measure the height of a room as accurately as by a rule; an air pump which exhausts perfectly – and of course is constructed on principles totally new – a balance which I have seen turn with [1/100th] of a grain, when loaded with a pound at each arm – I have all these instruments in the Elab[orator]y and several more of less importance – He is besides now taking out a patent for some new machinery which I believe will supersede all the water-wheels, steam engines &c now in use."[55] The "pastry cook" was the balloonist, mechanic, and engineer James Sadler, whose career was later frustrated by patent disputes with Boulton and Watt, and who was associated with Beddoes until the latter's departure from Oxford in 1793.[56] Beddoes's enthusiasm was always as immoderate as his depressions. 1/100th of a grain equals about 3×10^{-7} lbs, which means that Beddoes was claiming a sensitivity of one part in 300,000, very nearly equal to the best mechanical balances made in the eighteenth century. These were Lavoisier's great balance, made by Fortin and Mégnié, and the comparable balances made by Jesse Ramsden, the finest instrument maker in Britain, and by John Harrison, whose fame rests on inventing and building a chronometer that satisfied the Royal Navy's needs for determining longitude. These three balances could each weigh to about one part in 400,000.[57] No other instruments came near these three triumphs of precision craftsmanship. To put this in perspective, Joseph Black achieved his results with a balance sensitive to about one part in 200. Beddoes's claim here is simply not credible; nor is his account of a new barometer that could measure the height of a room. His gasometer, however, was effective. He sketched this instrument in his correspondence with Watt, who incorporated it into the breathing apparatus that he made for pneumatic medical purposes.[58] This gasometer used a weight over a pulley to control the pressure on the gas in the gasholder, on the same principle as those advertized by Dumotiez in Paris from 1795.[59] If one leaned towards simplicity and economy, improvements over Lavoisier's design were possible, and several chemists and instrument makers devised such simplifications and improvements. As for Sadler's steam engines, quite apart from the legal squabbles with Boulton and Watt, they were less successful than he and Beddoes wished to believe. The Shropshire ironmaster William Reynolds[60] had professional dealings with Boulton and Watt, making several engines for them. His company was sufficiently independent to give Sadler's machine a fair trial. The results were disappointing: "We have made no further progress in the trial of Sadlers Engine we are convinced it will answer no good purpose for large Engines." The report on the engine allowed that it could be useful for light work, for example in driving a lathe, but it would not serve for pumping out mines or driving large machinery.[61]

For all the problems with his engines, Sadler seems to have constructed a set of apparatus for Beddoes that enabled him to perform Lavoisier's key experiments, and to become persuaded that Priestley and the phlogistonists were wrong, and that on the whole the French chemists were right. I say "on the whole" because Beddoes was never fully persuaded that Lavoisier's list of elements was right, in detail and in some of its more general aspects. His early unhappiness about caloric was reinforced by Humphry Davy's experiments generating heat by friction, and also by Benjamin Thompson Count Rumford's better known experiments on generating a seemingly endless amount of heat by boring cannons. He also remained worried about the status of metals and of alkalis, until he had the excitement and satisfaction of learning about Davy's electrolyses of caustic soda and caustic potash, and the discovery of the new metals sodium and potassium. But long before Davy's *annus mirabilis*, Beddoes did unequivocally reject the phlogiston theory, towards the end of his teaching at Oxford.

NETWORKS IN SCIENCE, MEDICINE, POLITICS, AND INDUSTRY

Beddoes had to leave Oxford in 1793; he jumped before he was pushed, resigning once he learned that he would not be given the Regius Professorship that he had been all but promised. The reason, as he accurately surmised but was not told, was his public support for the French Revolution. Once he found his career in chemistry blocked at Oxford, he turned to the other profession for which he was qualified, that of medicine. He had been Black's pupil, and knew the importance of pneumatic chemistry; he had translated both Mayow and Scheele[62] from the Latin; he was fully aware of the importance of gases in the new French chemistry; and he had closely followed Priestley's experiments and critiques. He had been in correspondence with Erasmus Darwin since his return from France in 1787, a potentially useful connection, since Darwin was arguably the foremost physician in England, who more than once declined invitations to become the King's physician.[63]

As Reader in chemistry at Oxford, he had also become known to chemists at home and abroad. In 1791, travelling through Birmingham on a geologizing trip, he decided to visit James Keir, translator of Macquer's *Dictionnaire de Chimie*,[64] and an industrial chemist. "Fortunately he was at home. As our opinions in chemistry were different and in politics the same, only that I have scoured more of the rust of prejudice off my mind, and as he is the intimate friend of Darwin and Day, we should have been unlucky indeed if we had wanted conversation during the two days I passed with him ..." Thomas Day, who had died young in 1789, had been an author and political campaigner, an opponent of slavery, a supporter of the American Revolution, and, like Darwin and Keir, a member of the Lunar Society, the extraordinary group that also included James Watt, Matthew Boulton, and Joseph Priestley.[65] Keir had lately published a biography of his late lamented friend Day.[66] Keir and Beddoes talked about chemistry, about medicine, about Day, and about politics, and Keir showed Beddoes his factory, where he had made first caustic alkali, then soap, and red and white lead. As Beddoes was about to leave, Keir urged him to help with "his great chemical work, offering me any articles I might choose." Keir had published the first part of his dictionary of chemistry two years previously,[67] using his own copy of his translation of Macquer as the foundation,[68] and editing, removing, or adding material in the margins. Keir's first volume got through the A's as far as the article on "Affinity." Beddoes was not prepared to write an equal share of four or five volumes. "Besides he was the most able and conspicuous defender of the old system: to me the truth seemed to be on the opposite side." Keir himself must have been demoralized about his new dictionary, because the first part is all that ever appeared, and it did so in the same year that Lavoisier published his *Traité*. Beddoes for once made a wise decision.[69]

I do not yet know how Beddoes's intercourse with James Watt began, but a letter from Beddoes to Watt dated 8 January 1791 refers to a drawing of Joseph Black's furnace with which Watt had previously favored him.[70] They may have first been in touch about Sadler's engines, but Erasmus Darwin, a close friend of Watt's, is a likelier source. Thus by 1791, Beddoes was in touch with Darwin, Keir, Watt, and, at least as a customer, with Wedgwood, from whose factory he had ordered chemical apparatus.[71] When he left Oxford two years later, he kept in touch with them, and he shared with them the development of his ideas about pneumatic chemistry. Because of his political sympathies, Beddoes became especially close to James Watt, jr., son of the engineer. James Watt sr. wrote in exasperation to Joseph Black, his friend and mentor: "My Son James's conduct has given me much uneasiness, though I have nothing to accuse him of except being a violent Jacobin, that is bad enough in my eyes, who abhor democracy, as much as I do Tyranny, being in fact another sort of it." Knowing the Watts brought Beddoes into touch with the Boultons, father and son, and it was the son, Matthew Robinson Boulton, who facilitated Beddoes's acquisition of German books. Having already corresponded with the Wedgwoods about chemical apparatus, he soon found himself in regular touch with them. James Watt's first wife Margaret had died in childbirth in 1773. In 1776, Watt remarried, and he and Anne had two children, both of whom died of consumption: Janet (Jessy), who died in 1794, and Gregory, who died a decade later. Margaret Watt, James's daughter by his first marriage, died in 1791. Tom Wedgwood, youngest surviving son of Josiah and Sarah, also had serious health problems, including depression; he died in his mid-twenties, in 1805. The Watts and the Wedgwoods consulted local and not so local doctors, notably including their close friend Erasmus Darwin;[72] but their children failed to respond to a variety of treatments. Erasmus Darwin, James Watt, and Josiah Wedgwood all consulted Beddoes; Beddoes later treated Matthew Boulton, and Joseph Priestley's daughter Sarah, another victim of consumption.[73]

PNEUMATIC CHEMISTRY AND PNEUMATIC MEDICINE

Beddoes seemed to offer the Watts and Wedgwoods a lifeline. He was a physician and a chemist. At Edinburgh, he had made or witnessed experiments upon animals, to elucidate the action of air on the blood in its passage through the lungs. He pondered Priestley's and Lavoisier's ideas about respiration, and Black's teaching about airs, along with Guyton de Morveau's chemical approach to disinfecting air.[74] He was familiar with the role of the atmosphere in carrying disease in the form of "miasmata." He concluded that there was a connection between respiration and air, and saw this connection as essentially a chemical one. When Beddoes had to leave Oxford and turn to the practice of medicine, he fixed upon Bristol Hotwell because, as he explained in a public letter to Erasmus Darwin, "this resort of invalids seems more likely than any other situation to furnish patients in all the various gradations of consumption."[75] Looking back over medical records in Britain in the 1790s, Beddoes found from the mortality figures available to him that consumption was the leading killer disease. Of the 20,000 or so death records that he examined covering a period of two and a half years, roughly 5,000 a year were victims of consumption, as against 2,000 killed by fevers, and, on average, rather less than 2,000 killed by smallpox.[76] He had chosen well in going to Bristol.

Beddoes wanted to pursue the connection that he was sure existed between respiratory diseases and the chemistry of gases. This was the kind of project that appealed to Watt and Wedgwood, practical men with a knowledge of chemistry, some knowledge of medicine (Watt often acted as a physician to his friends and employees), intellectual enterprise, and a willingness to invest in their ideas and those of others. In 1793, soon after Beddoes settled in Bristol, Watt wrote to Black: "We have had no philosophical news since the affair of the frogs electricity except that Doctor Beddoes is applying the antiphlogistic chemistry to medicine, Azote and other poisonous airs to cure consumptions and oxigene for spasmodic asthmas he is at Bristol wells for the greater practice."[77] Maybe his work would help Jessy and Gregory Watt, as well as Tom Wedgwood. Beddoes was soon planning for a medical and chemical research institution, and Wedgwoods, Watts, and Darwins contributed handsomely, and encouraged others to contribute. In June 1794, Darwin wrote to Beddoes about the desirability of some account of the means to produce and administer the different gases.[78] Watt promptly invented a special apparatus for the preparation, storage, and therapeutic administration of gases; in July, Darwin wrote to him that it gave him "great satisfaction both on your account and on that of the public, that you are employing your mind on the subject of medicinal airs, of which indeed Dr. Beddoes had before informed me. You will do me a great favor by sending me an apparatus, or a description of one, and an account how easily to obtain the gases." Watt sent Darwin an apparatus later that month. Darwin thanked Watt for his "magnificent apparatus," and thought "both the joints, and the manner of dropping the water very ingenious, and truly Wattean." He suggested minor improvements, and urged Watt to write a pamphlet about the production and storage of different airs or gases.[79] Watt and Beddoes duly provided and published one. Watt's part of their joint publication showed how gases were generated in a special furnace, transmitted via a tube into a gasometer, and then administered to patients by means of another tube, an oiled silk bag, and a specially designed mouthpiece. The approach is similar to that taken by Lavoisier in his experiments on human respiration, but the apparatus was simpler, and Watt made small and large versions of his apparatus commercially available for use by hospitals, physicians, and even private individuals.[80] The enterprise that

Beddoes had initiated became well known in German states, but France at first was less aware of it. In May 1797, Beddoes wrote to James Watt: "I have had a certain account from Paris brought by a Bristol merchant lately arrived from thence that Fourcroy Morveau &c are very much interested by what we have attempted here on the subject and are likely to prosecute it – Is it not extraordinary that they did not know a word of the matter till Gimbernat who went from England told them though the German journals have been full of this business and more than one translation has been published of the considerations in German."[81] Gimbernat was a Catalan surgeon, and founder of the Royal College of Surgery of San Carlos, whose work on hernia surgery Beddoes had translated from Castilian in 1795.[82]

By 1798, Beddoes had his Pneumatic Institution, where he undertook the care of patients, using experimental techniques first tried on frogs, then on Davy and himself, his friends, and finally (but with much less delay than modern protocols would require) on hundreds of patients. I have elsewhere told the story of Beddoes's explorations in pneumatic medicine and chemistry, and of the successful campaign for the establishment of his Pneumatic Institution.[83] It is well known that Beddoes hired, as the chemical operator and then director of his chemical laboratory, a then unknown youth from Penzance, Humphry Davy. But until almost the last minute, Davy was not under consideration, except as a very junior assistant. At one stage, Beddoes was negotiating with Scherer, a German chemist whose pride appears to have been wounded by the offer of a subordinate position. If it was A. N. Scherer, a measure of pride would not have been unreasonable, since he had just started editing and producing one of the few chemical journals of the day. But a more likely candidate is Johann Baptist von Scherer, who wrote on chemical nomenclature, eudiometry, and Mayow, and would have been a good fit with (if not a rival to) Beddoes.[84] Then Beddoes started to negotiate with a young Irishman, "Mr. Boyd, at Dublin," recommended to him as being as good as William Higgins, who in Oxford had been operator to the Professor of Chemistry, William Austin, and may also have helped Beddoes when he became Reader. [85] Clearly there had been some unease about Boyd's competence, at least among the Watts, for in January 1798 he wrote to James Watt jr., enclosing a copy of the letter he had sent to the unfortunate Mr. Boyd. It is interesting in spelling out, in a deliberately daunting fashion, just what Beddoes expected of his chemical operator:

> It goes for nothing to be a speculatist: he must be a practical chemist & it is essential that he shd. not have lain principally in the line of pharmacy or any single line[.] [W]ith the more complicated exps. with gases he shd. be familiar; & also with those processess of fusion & calcination which require the reverbatory air and muffle furnaces. to sum up all he should be conversant with whatever illustrates the philosophy of chemistry as we find it in Lavoisier .... [3 words illegible] He shd. be master of the blow pipe – at least understand how to bend & manage the glass tubes, and <understand> the construction of furnaces. I should not think my self bound to terms with a person who did not answer to the above description liberally interpreted ... I should esteem it a singular mark of Candour if not having been conversant in the practice

> of experiment belonging to phi[loso]phical Chemistry, you should decline the proposal ... PS you would have excellent assistance I expect a young man who has applied close to chemistry (I mean worked hard) for 4 or 5 years he alone wd. be adequate in case of need to the situation of assistant.[86]

Boyd was no more heard from. Gregory Watt, meanwhile, boarding with Humphry Davy's widowed mother to take advantage of the gently healthful Cornish climate, had been impressed by her son, who was the young man who had applied himself to chemistry. Beddoes, impressed by Gregory Watt's account and that of Davies Giddy, hired Davy, a mere apothecary's apprentice. Beddoes wrote to James Watt: "I have been corresponding lately with Humphry Davy of Penzance, concerning whom apply to Gregory. I think him most admirably qualified to be the superintendent. I have read the account of some exps. of his; and he appears to me to have uncommon talents for philosophical investigations. He has besides entered with ardor into the career of chemical philosophy. Giddy entertains the same high opinion of his talents."[87] It turned out to be an inspired appointment, although Davy was for a while seduced by Beddoes's fertility in hypothesizing, and burned his own fingers by publishing speculations along with experiments.

By the end of the January, Davy was installed in the Pneumatic Institution, looking forward to the creation of a "superb laboratory," and giving his own advice to his friend Henry Penneck[88] in Penzance, to whom he sent apparatus. He told Penneck, "Chemical Implements are not kept ready for sale in the Bristol glass houses or I should have procured them long before. I was obliged to superintend the execution of them." He listed the apparatus sent, including a blow pipe, common retorts, one tabulated retort, two cylinders, a large glass receiver, a double necked receiver, curved tubes, glass tubes, matrasses,[89] and flint, "With these apparatuses: a few crucibles a small furnace and an Argand lamp you may make all the expts essential to Philosophic Chemistry." Davy stated ambitiously that he intended by the year's end "to publish a much larger work in the Laws of Corpuscular Motion and the connexion of chemistry with the Laws of life." Beddoes thought that was a reasonable goal for a brilliant neophyte.

CONCLUSION

Davy's ambition exceeded his reach, at least at this date. Nitrous oxide was found to be respirable, generally with no ill effects; Davies Giddy wondered whether gases that had the power "to diminish the secretion of excitability in the Brain may possibly be applied to many useful purposes. May it not be used before painful operations [?]"[90] At the end of the decade, Beddoes speculated that nitrous oxide "may probably be used with advantage in [operations in] which no great effusion of blood takes place."[91] But in spite of extensive use of nitrous oxide for therapeutic purposes in Beddoes's institution, and in hospitals and surgeries throughout England and Scotland, anesthesia was not seriously considered.[92] Margaret C. Jacob and Michael J. Sauter have argued that there were good reasons for this.[93] The virtues of nitrous oxide, carried by the blood, lay in reducing excitability; blood loss during surgery also reduced excitability, entailing a double loss of excitability, which could do serious harm. That argument makes sense within the framework of Brunonian medical theory, which Beddoes first adopted and soon came to criticize. There were also chemical, physiological, and mechanical issues that made the use of nitrous oxide as an anesthetic highly problematic.[94]

The Pneumatic Institution was intended to be a research institute devoted to determining the utility of different gases in treating different diseases, of which consumption was the most urgent. Beddoes thought that it would take only a few years to obtain the answers. He was right. The experiment showed that oxygen was helpful in several conditions, including asthma, if one ensured that the gas administered was pure. But no gas proved to be the cure for consumption, or for cancer, or for several other diseases about which Beddoes had once been optimistic. By 1801, he had his answers, and regarded the research program as a failure. He moved to a different address, where he managed a reformed institution providing medical assistance to the poor, with an emphasis on preventive medicine. The new institution rejoiced in the name of the Institution for the Sick and Drooping Poor.[95] Here, Beddoes's agenda was well conceived, but his declining health contributed to a growing depression about his entire career in medicine. In 1803, James Watt, who had lost his daughter Jessy and was soon to lose his son Gregory, still had some confidence in the medical powers of gases. He wrote to Beddoes: "You also seem to be among the skeptics in pneumatic medicine, you seem to have discarded both H. carbonate [methane] & oxygen which have done good here."[96] Oxygen no doubt did some good. It is also clear from the responses of patients that many of them were cured by breathing gases that were not in themselves beneficial; but belief in their medicinal qualities could generate cures, at least for some diseases, and at least in the short run. There is even a record of one patient, treated by Davy in the presence of Coleridge, who felt instantly better when a thermometer was placed under his tongue; whereupon Davy told him to come back at regular intervals, and repeated the performance with the thermometer until the patient, suffering from some kind of paralysis, was cured.[97]

In 1812, after Beddoes's early death and Davy's Bakerian lectures announcing his discovery of the alkali metals, Davy was knighted and married a wealthy bluestocking, Jane Apreece. Anna Maria, Beddoes's widow, reported that Davy was at a small party at Mrs. L. Ho[rner's], [her] ladyship was covered with diamonds. In the [course] of the evening Sr. H informed the company that he had had a dinner party at his house, he had given the air (gaseous Oxyd of azote) to a few friends – Mr. such a one, Lady L. the Duchess of – and looking modestly down with a careless air "I made the Princess of Wales take some."[98] Nitrous oxide had moved from pneumatic chemistry, through pneumatic medicine, to fashion.

ACKNOWLEDGMENT

The research for this chapter was supported by a grant from the Social Sciences and Humanities Research Council of Canada and by an award from the Secretaria de Estado de Universidades e Investigación, Spain. I am grateful to these bodies for their support. I wrote this paper at the University Pompeu Fabra, Barcelona, and thank Professor Antoní Malet warmly for his help and generous hospitality. I also wish to thank Professor Maria Morrás Ruiz-Falcó, Chair of the Department of Humanities there, who provided a privileged environment in which to pursue research.

NOTES

[1] Roy Porter, *Flesh in the Age of Reason* (London: Penguin, 2003), 413; Porter nevertheless wrote a medical biography of Beddoes (note 2).

[2] J. E. Stock, *Memoirs of the Life of Thomas Beddoes M.D.* (London, 1811); Dorothy Stansfield, *Thomas Beddoes M.D. 1760–1808* (Dordrecht: Reidel, 1984); Roy Porter, *Doctor of Society: Thomas Beddoes and the sick-trade in late-Enlightenment England* (London: Routledge, 1992). Larry Stewart and I, with the guidance of Hugh Torrens, are collaborating on a new biography in multiple contexts.

[3] Lazzaro Spallanzani, *Dissertations Relative to the Natural History of Animals and Vegetables*, trans. Thomas Beddoes, 2 vols. (London, 1784), translated from *Opuscoli di fisica animale e vegetabile, aggiuntevi alcune lettere relative ad essi opuscoli dal Signor Bonnet e da altri scritte all'autore*, 3 vols. (Venice, 1782).

[4] Torbern Bergmann, *A Dissertation on Elective Attractions, translated from the Latin by the translator of Spallanzani's dissertations* (London, Murray, and Edinburgh, Elliot, 1785), translated from "Disquisitio de attractionibus electivis" in vol. 2 of *Nova acta Regiæ societatis scientiarum upsaliensis* (1775). *The chemical essays of Charles-William Scheele. Translated from the Transactions of the Academy of Science at Stockholm, with additions by T. Beddoes* (London, 1786).

[5] Birmingham Record Office, James Watt Papers 4/44/55; Joseph Black to James Watt, 15 March 1780.

[6] See R. M. G. Anderson's entry on Joseph Black in the *Oxford Dictionary of National Biography*.

[7] Gloucestershire Record Office D303 C1/61; Beddoes to Trye, n.d. [winter 1784–85]. This letter is published in T. H. Levere and P. B. Wood, "Thomas Beddoes and the Edinburgh Medical School: A Letter to Charles Brandon Trye, c.1785," *University of Edinburgh Journal* 32, 1986, 36–39. For information about Trye, see the entry Charles Brandon Trye by D'A. Power, revised by Michael Bevan, in the *Oxford Dictionary of National Biography*.

[8] Stansfield, *Thomas Beddoes*, 33–34.

[9] Edinburgh University Library MS Gen 875/III/52, 53; Beddoes to Black, 6 November 1787.

[10] Birmingham Record Office, Boulton and Watt papers; Beddoes to Matthew Robinson Boulton, 15 July 1800.

[11] British Library, Sotheby 60 (2). This document is now missing; I have a microfilm made some years ago.

[12] Neil Vickers, *Coleridge and the Doctors 1795–1806* (Oxford: Clarendon Press, 2004), 37–78 is devoted to Beddoes's connection to Coleridge, and gives a good account of the former's medical theories and practice.

[13] Thomas Beddoes, *Über die Schwindsucht*, trans. Kuhn (Leipzig, 1803).

[14] *Essai sur le phlogistique, et sur la constitution des acides, traduit de l'anglois de M. Kirwan* [by Mme. Lavoisier]; *avec des notes de MM. de Morveau, Lavoisier, de la Place, Monge, Berthollet, & de Fourcroy* (Paris, 1788). This edition is discussed in Seymour Mauskopf, "Richard Kirwan's Phlogiston Theory: Its Success and Fate," *Ambix* 49, 2002, 185–205.

[15] Joseph Priestley, *Disquisitions Relating to Matter and Spirit. To which is added, the history of the philosophical doctrine concerning the origin of the soul, and the nature of matter; with its influence on Christianity, etc.* (London, 1777).

[16] Gavin de Beer, *The Sciences Were Never at War* (New York: Thomas Nelson, 1960); H. B. Carter, *Sir Joseph Banks 1743–1820* (London: British Museum, 1988).

[17] T. H. Levere, "Spreading the Chemical Revolution: The Dutch Connection," *Bulletin of the Scientific Instrument Society* 49, June 1996, 14–16.

[18] See W. A. Smeaton, "Louis Bernard Guyton de Morveau, F.R.S. (1737–1813) and his Relations with British Scientists," *Notes and Records of the Royal Society of London* 22, 1967, 113–30; Kirwan, *Essai sur le Phlogistique*.

[19] William A. Smeaton, *Fourcroy, Chemist and Revolutionary, 1755–1809* (Cambridge: W. Heffer, [1962]); also "French Scientists in the Shadow of the Guillotine: The Death Roll of 1792–1794," *Endeavour* NS 17, 1993, 60–63.

[20] Birmingham Record Office, Boulton and Watt papers; Thomas Beddoes to Matthew Robinson Boulton, 13 August 1793.

[21] Joseph DeBoffe, French bookseller (trading as J. C. DeBoffe) was at 7 Gerrard Street, Soho, London, from 1792–1807.

[22] Stansfield, *Thomas Beddoes*, 101.

[23] The list of Beddoes's Bristol borrowings has not been published. See G. Whalley, "The Bristol Library borrowings of Southey and Coleridge, 1793–98," *Library* 4 1949, 114–31.

[24] For a list of books and their reviewers, see Benjamin Christie Nangle, *The Monthly review, second series, 1790–1815; indexes of contributions and articles* (Oxford: Clarendon Press, 1955).

[25] Beddoes's reviews included one of Fourcroy, *La medicine éclairée par les sciences physiqus*, in *Monthly Review*, ser. 2, 12, 1793, 509; and Fourcroy, *Philosophie chimique* 18, 1795, 16.

[26] Smeaton, "Shadow of the Guillotine."

[27] Beddoes, *Memorial*, 15.

[28] Ibid., 19.

[29] Cornwall Record Office, MS DD DG 41/5; Beddoes to Giddy, 8 November 1792; Beddoes, *Hygëia, or Essays Moral and Medical on the Causes Affecting the Personal State of Our Middling and Affluent Classes*, 3 vols. (Bristol and London, 1802–03), 1:77, "As to the sort of reading, most injurious to young females, I cordially assent to the opinion of almost all men of reflection. NOVELS, undoubtedly, are the sort most injurious. Novels render the sensibility more diseased. And they increase indolence, the imaginary world indisposing those, who inhabit it in thought, to go abroad into the real."

[30] Beddoes to Trye.

[31] Joseph Black, *Lectures on the elements of chemistry, delivered in the University of Edinburgh*, John Robison, ed., 2 vols. (London and Edinburgh, 1803).

[32] See, e.g., William Cole, *Chemical literature, 1700–1860: A Bibliography with Annotations, Detailed Descriptions, Comparisons, and Locations* (New York: Mansell, 1988).

[33] See Paul Lawrence's essay on James Gregory, *Oxford Dictionary of National Biography*.

[34] Michael Neve in the *Oxford Dictionary of National Biography* gives the date of Beddoes's appointment as Reader as Spring 1788; but Beddoes styled himself Reader in his memorial on the state of the Bodleain Library, printed and issued in 1787. His lectures were not only on chemistry, but also on the closely related discipline of mineralogy, and on the theory of the earth, which in the late eighteenth century was enlivened by the conflict between the theories of Werner and Hutton.

[35] Edinburgh University Library MS Gen 875/III/52, 53; Beddoes to Black, 6 November 1787.

[36] Joseph Black, lecture notes taken by a student, Royal College of Physicians, Edinburgh, MSS M9. 42 p. 11.

[37] American Philosophical Society, MS BD315.1/1969.1821; Davy to Henry Penneck, 26 January 1798.

[38] Thomas Beddoes, ed., *Contributions to Physical and Medical Knowledge, principally from the West of England* (Bristol and London, 1799).

[39] Humphry Davy, "An Essay on Heat, Light and the combinations of Light," in ibid., 4–147.

[40] Alessandro Volta, "On the Electricity Excited by the mere Contact of Conducting Substances of Different Kinds," *Philosophical Transactions of the Royal Society of London* 90, 1800, 403–31.

[41] Beddoes in *Contributions*, 213.

[42] Black, *Lectures*, 1:193.

[43] For Davy's experiments and speculations, see David M. Knight, *Atoms and Elements: A Study of Theories of Matter in England in the Nineteenth Century* (London: Hutchinson, 1967).

[44] Edinburgh University Library, Gen 873/III/71, 72; Beddoes to Black, 23 February 1788.

[45] Ibid.

[46] Edinburgh University Library, Gen 873/III/129, 130; Beddoes to Black, 21 April 1789.
[47] Ibid. Beddoes to Black, 23 February 1788.
[48] Ibid.
[49] Keele University, Wedgwood archives, W/M 35; Beddoes to Thomas Wedgwood, 3 August 1793.
[50] Henry Richard Fox, afterwards Vassall, Baron Holland, *Further Memoirs of the Whig Party, 1807–1821, with some Miscellaneous Reminiscences*, Lord Stavordale, ed. (London: John Murray, 1905), 324.
[51] Lavoisier, *Traité élémentaire de chimie*, 2 vols. (Paris, 1789), 1:xxix–xxx.
[52] Beddoes to Black, 21 April 1789. This burning glass is the only piece of Beddoes's laboratory apparatus that has been unequivocally identified as his in the Museum of the History of Science in Oxford University, which once housed the university's chemical laboratory. The Museum also houses ceramic retorts made by Wedgwood, and these may have been acquired and used by Beddoes, who ordered apparatus from Wedgwood. See note 53 below.
[53] Beddoes to Josiah Wedgwood, quoted in R. E. Schofield, *The Lunar Society of Birmingham* (Oxford: Clarendon Press, 1963), 373.
[54] T. H. Levere, "Lavoisier's Gasometer and Others: Research, Control, and Dissemination," in *Lavoisier in Perspective*, ed. Marco Beretta (Munich: Deutsches Museum, 2005), 53–67.
[55] Edinburgh University Library, Gen 873/III/200, 201; Beddoes to Black, 15 April 1791.
[56] See Hugh Torrens's article on James Sadler, *Oxford Dictionary of National Biography*. See also Bodleian Library, Oxford, MS Dep. c.134/2; James Sadler to Thomas Beddoes, 14 January 1791, for Sadler's own statement of the case.
[57] John T. Stock, *Development of the Chemical Balance* (London: Her Majesty's Stationery Office, 1969); T. H. Levere, "Balance and Gasometer in Lavoisier's Chemical Revolution," in *Lavoisier et la Révolution Chimique: Actes du Colloque tenu à l'occasion du bicentenaire de la publication du 'Traité élémentaire de chimie' 1789*, ed. by M. Goupil with the collaboration of P. Bret and F. Masson (Palaiseau: ABIX-Ecole Polytechnique, 1992), 313–32.
[58] Birmingham Central Library, Watt papers MS 3219/4/27:9; Beddoes to James Watt, letter 1795 undated; Thomas Beddoes and James Watt, *Considerations on the Medicinal Use and Production of Factitious Airs*, part III (London and Bristol, 1795).
[59] Levere, "Lavoisier's Gasometer," 65.
[60] See Barrie Trinder's entry on William Reynolds in the *Oxford Dictionary of National Biography*.
[61] Cornwall Record Office, DD/DG 41.30; Joseph Reynolds, letter to Thomas Beddoes via Davies Giddy, 26 August 1791.
[62] John Mayow and Thomas Beddoes, *Extracts from 'Tractatus quinque medico-physici ... Studio Johannis Mayow' Oxford, 1674, trans. and ed. by Thomas Beddoes* (Oxford and London, 1790); *The Chemical Essays of Charles-William Scheele. Translated from the Transactions of the Academy of Science at Stockholm, with additions by T. Beddoes* (London, 1786).
[63] Desmond King-Hele, *Doctor of Revolution: The Life and Genius of Erasmus Darwin* (London: Faber, 1977); see also *The Letters of Erasmus Darwin*, ed. Desmond King-Hele (Cambridge: Cambridge University Press, 1981); a new edition is currently in preparation.
[64] Pierre Joseph Macquer, *A Dictionary of Chemistry... Translated from the French by James Keir... The second edition. To which is added, as an appendix, A treatise* [by James Keir] *on the various kinds of permanently elastic fluids, or gases* (London, 1777–79).
[65] Jenny Uglow, *The Lunar Men: Five Friends Whose Curiosity Changed the World* (New York: Farrar, Strauss and Giroux, 2002); Schofield, *Lunar Society*.
[66] James Keir, *An Account of the Life and Writings of T. Day* (London, 1791).
[67] James Keir, *The First Part of A Dictionary of Chemistry* (Birmingham, 1789).
[68] Keir's working copy is in the possession of the author.
[69] Beddoes's account of the visit to Keir is in a letter to Davies Giddy, 21 November 1791; Cornwall Record Office, MS DD/DG 41/48.
[70] Birmingham Record Office, MS 3219/4/43:5; Beddoes to James Watt, 8 January 1791.
[71] The Museum of the History of Science, Oxford University, has some ceramic retorts (inventory nos. 37798, 42340, 49504) and combustion tubes (47565, 44886, 52033) made by Wedgwood, dated to the early

nineteenth century; but dating such pieces is problematic, and it may be that one or more of these pieces was supplied to Beddoes ca. 1790.

72 Darwin lived in Derby, some 35 miles from Birmingham, but although he seldom attended meetings of the Lunar Society, he was regularly consulted by Watt and Wedgwood on medical matters.

73 Schofield, *Lunar Society*, 377; Schofield, *The Enlightened Joseph Priestley: A Study of his Life and Work from 1773 to 1804* (University Park, PA: Pennsylvania State University Press, 2004), 404.

74 W. A. Smeaton, "Platinum and Ground Glass: Some Innovations in Chemical Apparatus by Guyton de Morveau and Others," in F. L. Holmes and T. H. Levere, eds., *Apparatus and Experimentation in the History of Chemistry* (Cambridge, MA: MIT Press, 2000), 211–41.

75 Thomas Beddoes, *A Letter to Erasmus Darwin, M.D. on a new method of treating pulmonary consumption, and some other diseases hitherto found incurable* (Bristol, 1793), 40.

76 Thomas Beddoes, *Hygëia*, 2:5. Consumption was clearly something of a catch-all in the 1790s, but Beddoes recognized tuberculosis when he encountered it, as did Erasmus Darwin.

77 Birmingham Record Office, James Watt Papers MS 4/12/29; James Watt to Joseph Black, 17 July 1793.

78 Bodleian Library MS Dep. C134/2; Darwin to Beddoes, 6 February 1794, published in *Letters of Erasmus Darwin*, 94C.

79 Birmingham Record Office, MS 3219/4/28:6; Beddoes to Watt, 25 June 1794; MS 3219/4/28:36, Darwin to Watt, 3 July 1794, published in *Letters of Erasmus Darwin* 94J; Darwin to Beddoes, July 1794, in ibid. 94K; Darwin to Watt, 17 August 1794, in ibid., 94P.

80 Thomas Beddoes and James Watt, *Considerations on the Medicinal Use and on the Production of Factitious Airs*, parts I and II (Birmingham, 1794); part II is Watt's description of his apparatus. Watt and Beddoes, *Supplement to the description of a pneumatic apparatus, for preparing factitious air* (Birmingham, 1796), amplifies the description Watt gave in the former work.

81 Birmingham Record Office, MS 3219/4/29:13; Beddoes to James Watt, 26 May 1797.

82 Thomas Beddoes, ed. and trans., *A new method of operating for the femoral hernia. Translated from the Spanish of Don Antonio de Gimbernat,... To which are added, with plates by the translator, queries respecting a safer method of performing inoculation; and the treatment of certain fevers* (London: J. Johnson, 1795). See Acadèmia i Laboratori de Ciències Mèdiques de Catalunya, *Homenatge fet a Gimbernat per l'Universitat de Granada i a la cirurgia catalana pel Dr. Víctor Escribano* (Barcelona: Vda. Badia Cantenys, 1918). I am grateful to Professor Antoní Malet for this information, and for introducing me to the work of Núria Pérez Puig, who is carrying out research on Gimbernat.

83 T. H. Levere, "Dr. Thomas Beddoes: The Interaction of Pneumatic and Preventive Medicine with Chemistry," *Interdisciplinary Science Studies* 7, 1982, 137–47; "Dr. Thomas Beddoes and the Establishment of his Pneumatic Institution: A Tale of Three Presidents," *Notes and Records of the Royal Society of London* 32, 1977, 41–49. These and other essays on Beddoes are reprinted in Trevor H. Levere, *Chemists and Chemistry in Nature and Society 1770–1878* (Brookfield, VT: Variorum, 1994).

84 See J. R. Partington, *A History of Chemistry*, 4 vols. (London: Macmillan, 1962), 3:598.

85 Birmingham Record Office, Boulton and Watt papers; Beddoes to James Watt, 12 December 1797; A. N. Scherer, ed., *Allgemeines Journal der Chemie*, 10 vols. (Leipzig and Berlin, 1798–1803). For Higgins, see Stansfield, *Thomas Beddoes*, 17.

86 Birmingham Record Office, Boulton and Watt papers; Beddoes to James Watt jr., 2 January 1798.

87 See Birmingham Record Office MS 3219/4/29:32; Beddoes to Davies Giddy, 15 July 1798.

88 American Philosophical Society MS BD315.1/1969.1821; Davy to Henry Penneck, 26 January 1798.

89 Mattrass: A glass vessel with a round or oval body and a long neck, used by chemists for digesting and distilling. OED.

90 Cornwall Record Office MS DD DG 40/2; Davies Giddy to Beddoes, 7 January 1795. Giddy was writing about heavy inflammable air (given its numbing effects, probably carbon monoxide, which needed to be handled with great care – James Watt Jr. became unconscious merely sampling it).

91 Quoted by June Z. Fullmer, *Young Humphry Davy: The Making of an Experimental Chemist* (Philadelphia: American Philosophical Society, 2000), 222.

92 F. F. Cartwright, *The English Pioneers of Anaesthesia (Beddoes, Davy, and Hickman)* (Bristol: John Wright and Sons, 1952).

[93] Margaret C. Jacob and Michael J. Sauter, "Why did Humphry Davy and Associates Not Pursue the Pain-alleviating Effects of Nitrous Oxide?" *Journal of the History of Medicine and Allied Sciences* 57, 2002, 161–76.

[94] Neil Vickers, *Coleridge and the Doctors*, chs. 1 and 2, passim; Roy Porter, *Doctor of Society: Thomas Beddoes and the Sick Trade in Late Enlightenment England* (London: Routledge, 1992); Martin Wallen, *City of Health, Fields of Disease: Revolutions in the Poetry, Medicine, and Philosophy of Romanticism* (Burlington, VT: Ashgate, 2004); Levere, "Dr. Thomas Beddoes."

[95] Thomas Beddoes, *Rules of the Institution for the Sick and Drooping Poor* (Bristol, 1803).

[96] Birmingham Record Office, MS 3219/4/118:538; Watt to Beddoes, 6 May 1803.

[97] John Ayrton Paris, *The Life of Sir Humphry Davy, Bart. LL.D. Late President of the Royal Society, Foreign Associate of the Royal Institute of France, &c. &c. &c.*, 2 vols. (London, 1831), 1: 74–75.

[98] Bodleian Library Oxford, MS Dep. c135/1; Anna Maria Beddoes to Mrs. Edgeworth, 31 July 1812.

SEYMOUR MAUSKOPF

REFLECTIONS: "A LIKELY STORY"

> If we can furnish accounts no less likely than any other, we must be content, remembering that I who speak and you my judges are only human, and consequently it is fitting that we should, in these matters, accept the likely story and look for nothing further.[1]

HISTORY OF SCIENCE AS DRAMATIC NARRATIVE

Whether we admit it or not, historians of science, like all historians, are story-tellers. A good historical narrative – unlike most actual life experiences but like all good stories – has a dramatic structure. This includes a strong beginning, a highly charged, significant middle, which passes to a climactic, meaningful conclusion. The master dramatic narrative in the history of science is that of the Scientific Revolution, focused in astronomy and physics. A highly significant beginning (Copernicus) rises to a dramatic, meaningful, tension-filled middle in the lives and careers of Kepler and Galileo (notably Galileo's struggle with the Church) and passes on through Descartes and maybe Huygens to a satisfying climax and resolution in the synthesis of Newton.[2]

"Scientific revolutions" in general, particularly as constructed by T. S. Kuhn in *The Structure of Scientific Revolutions* share this dramatic narrative structure and, I would argue, this is in large part what makes Kuhn's work so compelling. A Kuhnian "revolution" has a beginning in a normal-science-producing paradigm; it rises to a highly charged dramatic middle through the accumulation of anomalies and the production of a "crisis"; paradigm-conflict then leads to a satisfying resolution in the replacement of an old paradigm by a new one. Kuhn's series of mini-drama revolutions proved to be an irresistible competitor to the much more bland sequence of endlessly progressive science.

Howsoever important science became in Enlightenment culture, the history of the eighteenth-century sciences has always posed problems of dramatic narrative. There is simply nothing to compare dramatically with the sweep of the Scientific Revolution or even with episodes of the nineteenth century, such as the Darwinian Revolution – with one exception: the Chemical Revolution, situated in the last quarter of the eighteenth century. Adumbrated in the nineteenth century,[3] the Chemical Revolution came to dominate historical discourse on eighteenth-century chemistry in the 1950s, at just the time the history of science itself was emerging as a professional discipline. It had not always been so hegemonic; the greatest historian of chemistry of the twentieth century, Hélène Metzger, had produced a classical study of early eighteenth-century chemistry in 1930.[4]

L. M. Principe (ed.), New Narratives in Eighteenth-Century Chemistry: Contributions from the First Francis Bacon Workshop, 21–23 April 2005, 177–193.

THE OVERTHROW OF THE PHLOGISTON THEORY

But in England,[5] France,[6] and above all, the United States, the Chemical Revolution, and its highly complex and dramatic hero, Antoine-Laurent Lavoisier, became the focus and *telos* of eighteenth-century chemistry scholarship. In the United States, the Chemical Revolution was given great dramatic punch by James Bryant Conant as "the overthrow of the phlogiston theory."[7] This was constituted as an epic duel between Lavoisier and his English adversary, Joseph Priestley. The "overthrow of the phlogiston theory" received iconic status in Kuhn's *Structure*, where it served virtually as the paradigm of a scientific revolution.[8]

The background to the narrative of the "overthrow of the phlogiston theory" lay in the explanation by Georg Ernst Stahl of combustion and calcination. Combustibles and metals were conceived of as compound substances composed of an earthy and an inflammable principle, the latter denominated as the celebrated (or notorious) phlogiston. The process of combustion was a decomposition reaction in which the combustible gives off phlogiston. The process of refining a metallic ore or "calx" by heating in the presence of charcoal to yield a metal (say, iron) was a chemical synthesis in which the ore absorbs phlogiston given off by the flammable charcoal to produce the metal.

The dramatic opening act of the overthrow of phlogiston was the set of experiments on combustion of sulfur and phosphorus carried out by Lavoisier in the late summer of 1772. Conducting the experiments on a "pedestal apparatus" over water but under a bell jar, Lavoisier noted that the aerial atmosphere was diminished during combustion. Moreover, by carefully comparing the weight of the experimental water after combustion with that of an equal volume of distilled water, Lavoisier was able to show that the product of combustion (acids that dissolved in the water) weighed more that the original weights of sulfur and phosphorus. Rather than a decomposition reaction, combustion appeared to be a synthesis of the combustion with "something" in the air to produce a heavier product.

The discovery of the "something" – "dephlogisticated air" – by Joseph Priestley two years later constitutes the second act of the drama. The third is Lavoisier's recognition of the role that this aerial fluid played in combustion, the calcinations of metals, and acid formation. By 1778, he had changed the name to "oxygen." The remaining acts involve Lavoisier's ringing polemic against phlogiston to the Académie Royale des Sciences in 1785, the demonstration that water was composed of two gases, one of which was oxygen, at about the same time, the winning over of the most progressive of the French chemists, the publication of an "elementary" textbook embodying the new "anti-phlogistic" chemistry and of a new scientific journal to proclaim it. Lavoisier's own tragic demise on the guillotine in the summer of 1794 during the Terror only adds to the dramatic power of the narrative. By the time of his death, his anti-phlogistic chemistry was well on its way to triumph.

Historiographical Consequences

The "overthrow of the phlogiston theory" makes a good, teachable narrative. But beyond that, much of the research on eighteenth-century chemistry in this country and abroad came to focus on it. This was particularly true of the school of research led by Henry Guerlac in the 1960s and '70s. He himself published a masterful account of the "first act" in his book, *Lavoisier – the Crucial Year*,[9] and his students – Gerry Gough, Carlton Perrin, and others – constituted something of a "Lavoisier industry" of research. This research expanded upon, refined, and even emended the Conant/Kuhn "overthrow of the phlogiston theory" narrative on some points, but the "overthrow of phlogiston theory" was itself by no means overthrown.[10]

Despite Metzger's earlier monumental study, the hegemony of the Chemical Revolution among modern historians of science threw into obscurity chemical activities during the earlier part of the eighteenth century except for those "ingredients" that fed into the narrative of the Chemical Revolution.[11] This obscurity was already signaled before the great outburst of Chemical Revolution scholarship by Herbert Butterfield in 1949. Invoking what he termed the "postponed scientific revolution in chemistry," Butterfield grappled with the question of why the "chemical revolution"[12] had taken place a century after the major part of the Scientific Revolution. He proposed to answer this by what he called "intellectual obstruction" to the recognition of the true nature of air and water. Stahl's phlogiston theory came in for particularly critical assessment[13] but Butterfield dismissed the early eighteenth century in general as a sterile interlude for chemistry:

> Although there appears to have been continued interest in chemistry and chemical experiments during the first half of the eighteenth century, it is perhaps true to say that no remarkable genius emerged to develop what had been achieved in the previous decades by Boyle, Hooke and Mayou.[14]

More recently (1985), and after a good deal of the Lavoisier industry had done its work, Thomas Hankins echoed Butterfield's denial of significance to early eighteenth-century chemistry in a gentler tone but with perhaps more devastating import. Hankins proposed that "the Chemical Revolution was more the creation of a new science than the change in an existing one. Before 1750, chemistry could not be regarded as an independent discipline."[15] In this assessment, Hankins was more concerned with the professional contexts of early eighteenth-century chemistry than with its scientific substance. But his account of pre-Lavoisier chemical developments was couched in terms of what had become the canonical background "ingredients" of the Chemical Revolution narrative: the development of pneumatic chemistry and of the phlogiston theory.

In fact, by the time Hankins wrote his disciplinary assessment of pre-Lavoisier chemistry, a counternarrative of the development of chemistry as a discipline had already appeared in Owen Hannaway's 1975 *The Chemists and the Word: The Didactic Origins of Chemistry*. As the subtitle indicated, Hannaway's narrative traced "the invention of chemistry as a discipline" back to the initiation of a didactic chemical textbook tradition by the humanist, Andreas Libavius, with his own textbook, the *Alchemia* of 1597.[16]

Hannaway's book was instantly recognized as a major study but his narrative did not quite achieve the hegemonic status that the Chemical Revolution enjoyed for late eighteenth-century chemistry. This was because the historiography of seventeenth-century chemistry (or "chymistry") was becoming a very complicated affair (with little dramatic narrative structure) by the 1980s with new research traditions, including seventeenth-century Paracelsianism and alchemy.[17] But, if no one made the case for the emergence of a chemical discipline with the force that Hannaway did, all of these historical research traditions implied distinct chemical (or alchemical) disciplinary developments.[18]

These research traditions were focused on the seventeenth century and earlier and had little directly to say about eighteenth-century chemistry prior to the Chemical Revolution. There was one ambitious attempt to provide a general narrative for aspects of eighteenth-century chemistry, and one that was alternative to "the overthrow of the phlogiston theory," namely Arnold Thackray's *Atoms and Powers*.[19] As its subtitle indicates, this was not so much an account of the disciplinary development of chemistry, or even of its substantive scientific development, but rather a comprehensive narrative of one important theme, the development of what Thackray took to be a Newtonian-based program to explore the microlevel of chemical reactivity, first in the dynamics of chemical affinity and then in the quantitative patterns of Daltonian chemical atomism. Thackray's own subsequent research moved from eighteenth-century to nineteenth-century topics and *Atoms and Powers* did not stem the rising hegemony of the "overthrow of phlogiston" focussed Chemical Revolution.

It is ironic that the first use of the term "the chemical revolution" as a book title in English was for a work that had nothing in common with Lavoisier and phlogiston. This was Archibald and Nan L. Clow's 1952 *The Chemical Revolution: A Contribution to Social Technology*, a book that dealt with the utility of chemistry in the First Industrial Revolution. Although the historical exploration of eighteenth-century chemistry and industry never ceased in the decades of the hegemony of Lavoisier/Chemical Revolution scholarship, this subject became very much subordinate to studies dealing with "purer" scientific themes.[20]

In summary then, the primary focus of scholarship on eighteenth-century chemistry in the period 1960–1990 was on the Chemical Revolution as constructed around the life, career, and achievements of Lavoisier. Except for selected topics that served as background for this construction (phlogiston theory, pneumatic chemistry), the chemistry of the first two-thirds of the eighteenth century was ignored or denigrated. Given the efflorescence of historical interest in seventeenth-century chemistry in this same period, the result was a major historiographical lacuna. Although the emerging perspectives on seventeenth-century chemistry were radically different from what they had been at mid-century, this scholarship made, if anything, even more insistent the problem of Butterfield's "postponed scientific revolution in chemistry."

A RECONSIDERATION OF EARLY EIGHTEENTH-CENTURY CHEMISTRY

The tempo of scholarship on Lavoisier and the Chemical Revolution increased in the late 1980s in connection with the commemoration of the bicentenaries of climactic events of that revolution: the new nomenclature of 1787 and the publications of Lavoisier's *Traité élémentaire de chimie* and of the first volume of the *Annales de chimie et de physique*, both in 1789. And yet, in these same years, also appeared the first comprehensive consideration of early eighteenth-century chemistry since Hélène Metzger. This was written by a scholar who, although arriving late on the scene, had become a major contributor to the Lavoisier industry: Frederic L. Holmes.

Holmes' research had focused more on the history of physiology than the history of chemistry. But one of his earliest publications had dealt with an important milestone in the history of chemical analytical techniques and, in 1985, he produced the most detailed study of the scientific career of Lavoisier that had yet appeared, but with a focus on Lavoisier's chemical physiology.[21] Nevertheless, all aspects of the canonic narrative of the Chemical Revolution were also addressed. Three years later, however, Holmes took up the subject of eighteenth-century pre-Lavoisier chemical research. But, rather than focusing on the standard background topics to the Chemical Revolution, Holmes turned his attention to research programs that had, *prima facie*, little to do with that late eighteenth-century episode. His opening salvo indicated the historiographical perspective he was going to take:

> Historians of science have found it difficult to view eighteenth century chemistry as anything other than the stage on which the drama of the chemical revolution was performed. So strong has the disposition been to identify the advent of the modern science with the chemical system established by Lavoisier between 1772 and 1789 that all earlier activity has been treated most often as a prologue to these climactic events.[22]

This, Holmes saw, led to "an essentially negative picture of eighteenth-century chemistry as a whole. The subject is identified more by what was missing than by what was present."[23]

Focusing on early eighteenth-century French chemistry (although with important consideration of German and Swedish chemical developments), Holmes highlighted both institutional and substantive components. The institutional component was the Paris Académie Royale des Sciences, reformed and reorganized in 1699. Although chemistry had been represented and practiced in the older Académie, the members of the reformed chemistry section established, in Holmes' view, the most progressive research programs in Europe. They included Nicolas and Louis Lemery, Étienne-François Geoffroy, and Wilhelm Homberg. Their work was informed by a new objective, one that marked the scientific work of the Académie generally: "the obligation to advance their science."[24]

The substantive feature was the array of research programs that these chemists pursued; the ones that Holmes discussed were the consideration and analysis of salts and of plant substances "in which eighteenth century chemists carried on a progressive investigative tradition, and in which they reached a major conceptual and methodological transformation independent of the questions central to the chemical revolution."[25] Holmes also gave some consideration to the development of eighteenth-century industrial chemistry. In the case of the development of the Leblanc process for the manufacture of soda, Holmes reiterated his point about the discounting of eighteenth-century chemistry not connected directly to the Chemical Revolution. Criticizing Charles

Gillispie's denial of any "scientific" influence on the invention of this process, Holmes gave an historiographical explanation for Gillispie's position:

> Assuming the canonical view that chemistry emerged as a theoretical science only with Lavoisier's theory of combustion, Gillispie looked for an influence of that theory on Leblanc and also found nothing. That he did not look for an application of a scientific discovery made fifty years earlier by a chemist in the Academy of Sciences reflects the scant attention that historians of science have paid to the investigative activity of early eighteenth century chemistry.[26]

In the next decade, the importance of the new style of chemistry carried out at the Académie Royale des Sciences in the early eighteenth century was given great emphasis by Ursula Klein, in one of the first works to attempt to bridge seventeenth- and eighteenth-century chemistry. In particular, Klein argued that a revolutionary change of perspective on the nature of chemical combination took place, especially by Geoffroy in his famous publication, the "Table des différent rapports observés en Chimie entre différents substances." Combination came to be seen as the union of distinct, tangible chemical species, whose perseverance in the compound meant that they could be recovered by means of analysis. The new concept of chemical combination supplanted the seventeenth-century notion that chemical operations involved the extraction of "essences" from a material matrix by means of distillation.[27] Holmes' and Klein's emphasis on the significance of early eighteenth-century French academic chemistry has recently been elaborated on by Mi Gyung Kim, who developed a narrative that encompasses the entire eighteenth century.[28]

Mention should also be made of a number of works that have examined eighteenth-century institutional contexts of chemistry, although without any agenda of reinforcing or supplanting the traditional focus on the Chemical Revolution. The two most important are Karl Hufbauer's *The Formation of the German Chemical Community, 1720–1795*[29] and Arthur Donovan's *Philosophical Chemistry in the Scottish Enlightenment: The Doctrines and Discoveries of William Cullen and Joseph Black*.[30]

If Holmes, Klein and Kim have not produced a new dramatic narrative for all of eighteenth-century chemistry, each has certainly suggested important directions for emending and expanding one beyond the *telos* of the Chemical Revolution.[31] This workshop continued these recent efforts to reconfigure eighteenth-century chemistry. Indeed, perhaps the one theme common to the scholarly papers presented was the *avoidance* of consideration of the hitherto dominant episode of the Chemical Revolution.[32] While it would have been overly sanguine to expect a new, coherent master narrative to emerge from a series of discrete, independently produced research papers, still the papers of this volume do suggest further directions for reconfiguring eighteenth-century chemistry. Indeed, they are particularly significant for addressing topics and themes that have been largely ignored or suppressed in the Chemical Revolution narrative. Whether or not an alternative narrative emerges, the material taken up at this workshop will have to be incorporated in future narratives on eighteenth-century chemistry.

The papers of the workshop fall into two chronological and thematic divisions, corresponding to the order of presentation. Those presented in the first part of the workshop (the Francis Bacon Lecture by Lawrence Principe and the papers of the first day) have something of a common focus, one that elaborates on a theme developed by Holmes, Klein and Kim. This is early eighteenth-century *origins*, particularly, the establishment of chemistry as *a species of rational knowledge*. For the workshop papers, this entailed two particular components: the clarification of the disciplinary domain of chemistry (particularly its delimitation from alchemy) and the development of a "rational" methodology. These origins are broadened out to include not only French chemists of the Académie Royale des Sciences (as had largely been the case with the above-mentioned authors) but also two major non-French figures: Hermann Boerhaave and Georg Ernst Stahl. The papers of the second day are more difficult to encapsulate because they may seem to constitute an eclectic group. But, since their time frame falls later in the eighteenth century than the first group of papers, it is particularly in this group of papers that themes and topics largely absent from the narrative of the Chemical Revolution surface. Moreover there *is* a governing theme present in all of these papers: the role of chemical *application* in the practice and development of chemistry. How this theme is developed in different contexts and countries will be explored in my discussion of these papers.

ORIGINS: CHEMISTRY AS RATIONAL NATURAL KNOWLEDGE

"Reason" was, of course, a central trope to eighteenth-century rhetoric but it is a very complex and slippery term. What it meant for chemistry and why and how chemistry became rational at this time were questions taken up in the opening Bacon Lecture of Principe and both of the papers on Boerhaave; it is likewise implicit in Chang's paper on the abjuring of alchemy by Stahl. What emerges from this set of papers is that there was no one governing program to make chemistry rational; rather, moves towards rational chemistry appear to have been responses to local conditions (institutional and cultural) in which chemists found themselves. However, there seems to have been a common incentive behind these responses: chemists were concerned to ameliorate the perceived uncertain (and vulnerable) position of chemistry vis-à-vis other natural philosophical enterprises with which they were associated (the Académie) or which they serviced (Dutch and German medical faculties). Because of the local contexts, the construction of "rational chemistry" meant different things in different locales.

The chemists associated with the Académie, the focus of Principe's Bacon Lecture, are now well-known, to the *cognoscenti* at least, as pivotal to the transformation of the chemical research agenda for the eighteenth century. Principe deals with their activity in a related enterprise: the refashioning of the image of chemistry into one worthy of its position in the Académie. The challenge for such an image makeover came from social and political circumstances that had tarnished the image of the "alchemical" enterprise (scandal, satire, and rumor of nefarious practices) and from attitudes within the Académie itself. Here, it was derogatory view of chemistry (and alchemy) held by the Académie's most commanding figure, the perpetual secretary, Bernard de Fontenelle. To Fontenelle, chemistry, in contrast to *la phisique*, was confused and obscure, and he was far more contemptuous of transmutational alchemy.

The refashioning of chemistry's image involved two major moves: the sharp separation of chemistry from alchemy (with the denigration of the latter) and the imposition on chemistry of a more elevated and appropriate philosophical genealogy, involving Descartes, Boyle and, eventually, Newton. The move to separate alchemy from chemistry was supported by many – but not all – of the academic chemists. Indeed, the most important member of that group, Wilhelm Homberg, remained an alchemical adept throughout his life and his colleagues may have felt obliged to suppress his papers and unpublished writings after his death.

The move to elevate chemistry philosophically appears to have been fostered primarily by Fontenelle, but, according to Principe, it took on a life of its own over the next three centuries. Indeed, the principal burden of Principe's paper is to carry out an historiographical cleansing of the purported philosophical influences on seventeenth- and eighteenth-century chemists to reveal what were really chemical ones. It will be interesting to see what kind of debate ensues over Principe's claim. Behind Principe's arguments lie important perspectives on the nature of chemistry and its history: chemistry is not just watered-down physics but a distinct science in its own right with special characteristics. Unlike physics, which deals with universal properties of matter, chemistry studies the variety and diversity of material substances and their transformations as expressed by their complex sensual properties. Secondly, chemistry is as much concerned with fabricating materials as it is with determining their scientific nature. Hence, chemists throughout the eighteenth century were often craftsmen as well as natural philosophers – most of the early chemical *academiciens* were apothecaries – or were concerned professionally with craft supervision. This mixture of craft occupation or oversight with philosophical and academic status resulted in an uncertain relationship between chemists and other academic natural philosophers.

Such unease was the context behind the "rationalization" of chemistry for Herman Boerhaave, the Dutch contemporary of the Parisian academic chemists, although rationalization in this case meant something quite different from the French construction. To John Powers and Rina Knoeff, it signified the adoption of a theoretical skepticism and experimental empiricism modeled after the medical practice of the ancient Greek physician Hippocrates. Given Boerhaave's primary medical orientation, one might expect this to have involved nothing more than a rather uncomplicated transfer of medical methodology to chemistry.

Powers underscores such a transfer but develops a context for it in the problematic status of chemistry at Leiden University where Boerhaave taught. Chemistry as a pedagogical subject had entered European medical faculties in the seventeenth century but existed in an uneasy, subordinate position. At least, this seems to be how Boerhaave perceived the situation at Leiden ca. 1700. The prevalent ideology in the Leiden medical faculty was Hippocratic, and Powers argues that Boerhaave's appropriation of a Hippocratic stance for chemistry was motivated by his desire to preserve chemistry and enhance its status as part of medical teaching. Powers extends his discussion beyond general methodological considerations to substantive ones, particularly the contexts for Boerhaave's formulation of the concept of chemical "instruments" as a response to the skepticism towards the existence and perseverance of elements and principles in chemical substances expressed by another of Boerhaave's heroes, Robert Boyle.

Rina Knoeff agrees with Powers about the Hippocratic empiricist orientation of Boerhaave's chemistry but she explores a quite different context for it: a religious one residing in Boerhaave's Calvinism. Two points are especially pertinent to her explanation for Boerhaave's chemical empiricism. Boerhaave saw Hippocrates as especially concordant with a Calvinist religious conception of Nature because of a common providentialism. But, beyond this, the Calvinist contrast between divine omniscience and human intellectual limitations led Boerhaave to reject ambitious theoretical attempts to encompass all natural phenomena (such as Cartesianism and Newtonianism) in favor of what was possible for the human mind: experimental empiricism. Of course, there is no reason why Powers' and Knoeff's explanations for Boerhaave's version of rational chemistry have to be mutually exclusive. It is, indeed, more than likely that a variety of factors were operating in each of these examples to bring about the transformation of chemistry.

At this point, I want to express my delight over the extensive historical treatment given to Boerhaave at this workshop. He has been another casualty of the development of a canonical master narrative of the Chemical Revolution; despite his significance for eighteenth-century chemists and for Lavoisier, he has essentially been eliminated from this narrative. Powers provides some explanation for this in terms of the puzzlement and even hostility of the constructors of this narrative to Boerhaave's instrumental matter theory.

In contrast to Boerhaave, Georg Ernst Stahl has always been a major protagonist in the Chemical Revolution narrative. But, perhaps because his role has been essentially that of a "fall guy," his chemistry has not received nearly the historical consideration it deserves.

Kevin Chang, who has been rectifying this omission, here discusses an aspect of Stahl's abandonment of his earlier support of alchemy by the early 1720s. In contrast to the papers already discussed, Chang's focus is not on the change in Stahl's personal view *per se*; he asserts that Stahl "left behind almost nothing that documents the personal or social factors that occasioned this shift." Nevertheless, Chang's paper is, in a sense, biographical: he is trying to clear up serious confusion in the historical literature about when Stahl actually turned against transmutational alchemy. His strategy is to unmask what he terms an "apparent anachronism" caused by the printing of early works of Stahl supporting transmutational alchemy (including the ironically entitled *Chymia rationalis et experimentalis*) by opportunistic publishers at the very time when Stahl was becoming a strong skeptic.

It is unfortunate that no personal record has survived of Stahl's change of attitude. Was Stahl, like the other early eighteenth-century European leaders so far discussed, attempting to remediate a perceived vulnerability in the status of chemistry? Was he, perhaps, even influenced by the denigration of alchemy taking place among some contemporary French academicians? The evidence Chang gives from Stahl's 1726 "Bedencken von der Gold-Macherey" is suggestive. Paralleling French developments, Stahl distinguished between *Chymie* and *Alchymie*, the former rational and the latter now incredible, fraudulent and, ultimately, associated with disturbance of the social and economic order. Moreover, Stahl appears to have based his criticism of alchemical metallurgy and transmutation on the new chemical notion of perseverance of chemical species in chemical change, a change that Ursula Klein has documented as arising in the early eighteenth century, especially with Geoffroy.

In the absence of concrete personal testimony, all this must remain speculative. But, while I am indulging in *ex silentio* speculations, I'll make an even more ambitious one. The temporal juxtaposition of Stahl's change-of heart towards alchemy and the publication of his early alchemical works *may* relate to a broader European cultural change taking place at this time: a bifurcation in natural knowledge, along social lines, between elite (and professional) "rational" knowledge, and popular interest in discredited knowledge like alchemy, astrology, as well as the various religious "enthusiasms."[33] The bifurcation between elite-rational and vulgar-discredited knowledge might be worth exploring further as context for the other cases of the other early eighteenth-century chemists examined here. They, after all, were clearly responding to a felt uncertainty about the position of chemistry in the university and the academy, an uncertainty with social and cognitive bases going back at least to Boyle's denigration of "sooty empirics."[34]

Regarding the publication of Stahl's early alchemy-friendly works in the 1720s, Chang suggests that audience expectation was very great on the part of the publishers, especially Roth-Scholtz. It would be interesting to know how the extent and nature of the expected audiences for Stahl's German-language publications in medicine and chemistry compared to those for the alchemical publications.

CHEMISTRY AS USEFUL PRACTICE

As Principe has already made clear, the uncertainty in the status of chemistry derived in large part from its close links with craft productions and commercial interests. Chemistry has always been concerned with fabricating materials: refining, perfecting and deploying ones already known for useful purposes and, certainly by the eighteenth century, discovering new ones. Precisely the emphasis on fabrication (rather than theoretical explication) was a factor that often caused problems in chemistry's status relative to other sciences. Its mixed nature as craft-like and theoretical made it the quintessential *applied* science. Yet there was another, positive, side of this coin for the eighteenth century. The notion that scientific knowledge would be useful was perhaps second only its denomination as rational among eighteenth-century tropes. No area of natural knowledge was better suited to fulfill this promise than chemistry.

As is well known, chemistry has always been intimately associated with a variety of craft and commercial interests: mining and metallurgy, drug-making and the production of medical remedies, ceramics, painting and dyeing, to name a few.[35] Starting in the seventeenth century, chemistry became an established part of academic medical curricula[36] and, in the eighteenth century, many of the chemically oriented crafts came under state regulation and the craftsmen themselves were increasingly expected to have some training in experimental techniques. Moreover, new domains of chemical application emerged in the eighteenth century in connection with the Enlightenment ideology of social and economic improvement through the utilization of natural knowledge. The two most significant ones were industry and agriculture, as Karl Marx famously noted a century later in his *Communist Manifesto*.[37]

Yet these social and technological dimensions of the chemical enterprise were largely discounted with the hegemony of the Chemical Revolution narrative. The process is strikingly illustrated in the publications of the principal constructor of this narrative, Henry Guerlac. In 1959, Guerlac published an article with a tantalizing title: "Some French Antecedents of the Chemical Revolution."[38] The "French antecedents" turned out to be technological transfers from Germany to France in the form of French translations of German treatises on mining and metallurgy that highlighted the phlogiston theory among other matters. But this promising avenue of research was soon supplanted two years later by Guerlac's detailed focus on Lavoisier in *Lavoisier – the Crucial Year*. The "French antecedents" had almost – but not quite – dropped out of his account. In the preface to this book, Guerlac criticized the characterization of the Chemical Revolution as being equivalent to the overthrow of the phlogiston theory although he still centered it in the work of Lavoisier. What he saw as "the most significant [and overlooked] ingredient" in it was that "at his [Lavoisier's] hands, the pharmaceutical, mineral, and analytical chemistry of the Continent was fruitfully combined with the results of the British 'pneumatic' chemists who discovered and characterized the more familiar permanent gases."[39] But the so-called Continental chemical enterprises, along with those of mining and metallurgy, were not pursued further by him.

Yet despite the hegemony of the Chemical Revolution narrative, research into eighteenth-century chemical application continued during the decades of its sway. But most of it has dealt with the context of the development of industrial chemistry, not surprisingly since this century witnessed the onset of the Industrial Revolution. This had been true of Clows' *Chemical Revolution*, and was largely the case regarding studies of applied chemistry both in Great Britain and even in France.[40] The papers in this workshop dealing with chemical application focus, however, on other domains. The two papers with a British venue (Eddy's and Levere's) deal respectively with agriculture and medicine. Through the activities of his protagonist, James Anderson, Matthew Eddy shows how Scottish academic chemical research and teaching on lime by William Cullen and Joseph Black were translated and transmuted into popular utilitarian forms for "georgics" or agricultural improvement. Trevor Levere's study of Thomas Beddoes has a wider purview because Beddoes, a more serious chemist than Anderson, was an interesting early "receiver" of Lavoisian anti-phlogistic chemistry. Yet the parallels of Beddoes' deployment of the discoveries in pneumatic chemistry for medical therapy in his Pneumatic Institution with Anderson's "georgics" are obvious. Moreover, Beddoes was closely associated with (and received patronage and support from) some of the leading Midland industrialist members of the Lunar Society, such as James Watt.

The trajectories of the fates of these chemical applications were quite different. Beddoes soon realized that gases were not the panacea he had hoped they might be. Lime (and limestone), on the other hand, if not quite the agricultural "philosophers' stone" touted by Anderson, remained an important component of soil amelioration and figured importantly in Humphry Davy's *Elements of Agricultural Chemistry*. Studies such as Eddy's and Levere's create a more historically realistic landscape of applied chemistry in eighteenth-century Britain than the concentration of industrial application alone provides. European society was still predominantly agricultural, even in Great Britain, and chemistry as an institutionalized science was associated predominantly with medicine and pharmacy throughout the century.

The final pair of papers, by Ursula Klein and by Bernadette Bensaude-Vincent and Christine Lehman, are oriented around another area of application: pharmacy. But, unlike Eddy and Levere, who study the process of transference (and transformation) of chemical knowledge to application, Klein and Bensaude-Vincent and Lehman focus on social and cultural features of the craft itself: Klein on the tradition of "apothecary-chemists" in Germany, and Bensaude-Vincent and Lehman on the function of eighteenth-century French public lectures on chemistry in training apothecaries and physicians.

Klein not only resurrects a marginalized chemically-oriented occupation but, equally significant, an extremely important geographical area for the pursuit of chemical knowledge – Germany – that had been all but eliminated from the Chemical Revolution narrative. Between Stahl (and phlogiston) and the reception of anti-phlogistic chemistry in the 1790s, German-speaking chemists are largely absent from this narrative.[41] Yet the German apothecary-chemist tradition produced some of the most celebrated European analytical chemists of the century. Two of those discussed by Klein – Marggraf and Klaproth – were among the most prolific discoverers of new chemical substances. Adding Karl Wilhelm Scheele, another apothecary-chemist, to this group would make it indeed the most formidable discovery-engine of the later eighteenth century. Yet the role and achievements of German apothecary-chemists, although certainly known to historians of chemistry, have not been incorporated into a narrative of mid- and late-eighteenth-century chemistry.[42]

It would be most interesting to know in more detail what it was about their training that facilitated such impressive technique in chemical analysis. A general cultural factor that undoubtedly contributed to the formation of the community of apothecary-chemists and, indeed, to its success was the very high literacy of the eighteenth-century German population. In case after case, apothecaries were able to supplement, or rectify, their apprentice-training through the perusal of chemical texts and, later, of journals.

The apothecary-chemist tradition in France has also lain hidden under the shadow of the Chemical Revolution narrative. The presence of apothecaries and iatrochemical physicians in seventeenth- and eighteenth-century chemistry has certainly been known to historians of chemistry but, until very recently, their role in the development of the French chemical enterprise has not received anything like its due recognition. Like Klein, Bensaude-Vincent and Lehman consider the technical formation of apothecaries but their focus is on how chemical knowledge was imparted through the medium of public lectures rather than through apprentice training. Because of the nature of this genre of instruction, Bensaude-Vincent and Lehman necessarily broaden socially their discussion of chemical pedagogy beyond technical knowledge *per se*. Lectures whose functions included both the imparting of technical knowledge to specialized groups *and* theatrical entertainment (and edification) to a larger public represent a peculiar (and particular) stage in the pre-specialization of knowledge and the pre-institutionalization of technical teaching. Although they had their share of showy sensationalism (for which Rouelle was famous), the lectures in France and Britain, at least, were basically the purveyors of rational and useful knowledge to the public audiences. What is most striking about them – and about their itinerant English counterparts, such as those of Peter Shaw and William Lewis – is the prominence of chemistry.

As suggested by Bensaude-Vincent and Lehman, the theme of chemistry as rational popular knowledge could be pursued further through the exploration of the role of chemistry in the provincial societies and academies. These were burgeoning throughout Europe in the second half of the century; of them, the Dijon Academy and the Manchester Literary and Philosophical Society were the most celebrated vis-à-vis chemistry. But even at the European scientific periphery, chemistry was playing an important role in provincial societies. One example that I have touched on in my own research involves the career in Spain of Joseph-Louis Proust. The young Proust was first invited to Spain by the *Sociedad Bascongada de los Amigos del Pais* to teach in the *Sociedad's* school, the *Real Seminario de Bergara*, in 1778. He spent most of the next 30 years in Spain teaching and lecturing in various institutional and public venues.

All of these activities concern chemistry as rational knowledge. How did the discredited alchemical knowledge fare? Allen Debus has shown that publications in alchemy actually peaked in the eighteenth century. In the early nineteenth century, Mary Shelley testified to the continued popularity of alchemy in the intellectual trajectory of her hero, Victor Frankenstein. It will be recalled that the young Frankenstein's first introduction to natural philosophy was through the works of Cornelius Agrippa and Paracelsus. Whether or not the continued dissemination and popularity of alchemy and its transformation into a topic classed alongside other semi-secret, sensationalist, "occult" interests is deemed a legitimate part of a master narrative on eighteenth-century chemistry, it certainly deserves serious historical attention and further exploration.

CONCLUSION

> À la fin du XVIIe siècle, la chimie apparaît comme une discipline au sens précis de "matière enseignée," et cette matière est, dans sa majeure partie, étroitement solidaire de la médecine, et de pratiques artisanales telles que la métallurgie, la parfumerie, etc. À la fin du XVIIIe siècle, la chimie est reconnue comme une science à part entière, autonome, légitimite, assise sure des bases solides et source d'applications utiles au bien public. Comment a-t-elle conquis ce statut?[43]

Aside from its riveting dramatic structure, the Chemical Revolution narrative rhetorically validates chemistry as an academic science whose principal objective is discovering principles of natural knowledge rather than fabricating useful things (even if still full of useful promise). This is, above all, the *telos* of the Chemical Revolution narrative and explains, I believe, its great attraction to Thomas Kuhn, who wanted to see the commonalities between the various sciences rather than the differences. However, as we have seen, this narrative eliminates an equally valid side of chemistry: chemistry as craft identity and craft knowledge utilized for material productions. The validation of chemistry as a science in the above sense fitted in well with structural changes taking place in French science in the late eighteenth century and, more generally, in European science in the next century: the emergence of professional, research-oriented, specialized academic science. It is surely no coincidence that the construction of the Chemical Revolution first clearly emerged at this time.[44]

Does this mean that the Chemical Revolution narrative was (and is) simply a legitimating device that can be dispensed with, now that we have unmasked its true function? Can we imagine an alternative narrative for eighteenth-century chemistry with a different *telos*? Holmes, who pioneered the broadening out of historical consideration of chemical research beyond (and before) that of the Chemical Revolution, himself had second thoughts about whether the Chemical Revolution might not indeed still be the climax and *telos* of the narrative of eighteenth-century chemistry.

I must confess that it is not clear to me whether an alternative dramatic narrative to the Chemical Revolution for the range of eighteenth-century chemistry (as outlined here) will be possible. But let me suggest a general scheme. Its basis lies in the consideration of the unique nature of chemistry – and certainly eighteenth-century chemistry – as encompassing the rational and the utilitarian, the academy and the workshop. The first set of papers, dealing with what I termed "origins," dealt with the first major moves to emphasize the academic aspect of chemistry; one might even consider them to be precursors of what the Chemical Revolution itself accomplished. Yet the apothecary and physician protagonists of this first chemical "revolution" (Principe) were themselves exemplars of the craft traditions. And, throughout the eighteenth century, the academic and the craft side of chemistry coexisted, often in the same persons (as Rouelle, Macquer, and the German apothecary-chemists illustrate). More generally, the utilitarian side of chemistry was never far from even the most "academic" chemist, as Lavoisier's work in munitions testifies.

One way of bringing these different aspects of eighteenth-century chemistry into harmony may lie in revisiting Henry Guerlac's insight that the basis of the Chemical Revolution lay in the melding of British pneumatic chemistry with the Continental enterprises of pharmaceutical and analytical chemistry. As I noted, Guerlac himself never followed it up; perhaps it is time for us to do so. As some of the papers in this conference testify, our comprehension of "pharmaceutical and analytical chemistry" has both vastly expanded and become much more concrete in the ensuing 45 years. We are in a better position to define and delineate what these enterprises were, and to explore how they intersected and interacted with the research programs of Lavoisier and his colleagues. Whether or not a narrative as coherent and compelling as "the overthrow of the phlogiston theory" will result, it will certainly be a richer one, and one more historically faithful to the multifaceted nature of eighteenth-century chemistry.

NOTES

1 Plato, *Timaeus*, trans. Francis M. Conford (Indianapolis, IN: Bobbs-Merrill, 1959), 29:d (18).

2 The dramatic narrative works less well in the biomedical sciences, not well at all in the natural historical sciences of this period, and even less well in the chemical sciences. Significantly, constructions of the narrative of the Scientific Revolution have never known what to do with Paracelsus. The historian of science who has worked most assiduously to include Paracelsus and the Paracelsians in the Scientific Revolution narrative is Allen G. Debus. "Chemists, Physicians, and Changing Perspectives on the Scientific Revolution," *Isis* 89, 1998, 66–81.

3 See Mi Gyung Kim's superb account of its construction by nineteenth-century French chemists: "Lavoisier, the Father of Modern Chemistry?" in *Lavoisier in Perspective*, ed. Marco Beretta (Munich: Deutsches Museum, 2005), 167–91.

4 *Newton, Stahl, Boerhaave et la doctrine chimique* (Paris: Alcan, 1930). An early, classic history of chemistry that gave due weight to other factors and traditions in eighteenth-century chemistry while also recognizing the importance of the work of Lavoisier and anti-phlogistic chemistry is Ernst von Meyer, *A History of Chemistry from Earliest Times to the Present Day*, trans. George McGowan (London: Macmillan, 1898).

5 Douglas McKie, *Antoine Lavoisier, The Father of Modern Chemistry* (London: V. Gollancz, Ltd., 1935) and, more pertinently, *Antoine Lavoisier, Scientist, Economist, Social Reformer* (New York: H. Schuman, 1952); J. R. Partington, *A Short History of Chemistry* (New York: Macmillan, 1937). An American history of the period, *The Historical Background of Chemistry*, by Henry M. Leicester (New York: John Wiley & Sons, 1965; first edition 1956) is somewhat different in that it treats both the phlogiston theory and the theory of affinity as eighteenth-century background.

6 Maurice Daumas, *Lavoisier théoricien et expérimentateur* (Paris: Presses universitaires de France, 1955).

7 "Case 2: The Overthrow of the Phlogiston Theory: The Chemical Revolution of 1775–1789," in *Harvard Case Histories in Experimental Sciences*, 2 vols. (Cambridge: Harvard University Press, 1957), 1: 65–115.

8 Thomas S. Kuhn, *The Structure of Scientific Revolutions*, 2nd ed. (Chicago: University of Chicago Press, 1970), 53–60, 69–72, 86, 118 and passim.

9 Henry Guerlac, *Lavoisier – the Crucial Year: The Background and Origin of His First Experiments on Combustion in 1772* (Ithaca, NY: Cornell University Press, 1961).

10 Guerlac himself expressed reservations about the "overthrow of the phlogiston theory" narrative in the introduction to *Lavoisier – the Crucial Year*, xvii: "An equally common appraisal of the Chemical Revolution makes it tantamount to the overthrow of the Becher–Stahl phlogistic theory of combustion. But this says at once too much and too little; it exaggerates the break with the past; it neglects the accumulated body of old and recent factual knowledge that was absorbed unaltered by the newer chemistry; and it overlooks the point that something more fundamental occurred than the mere substitution of one theory of combustion for another, centrally important though this proved to be."

Some of the suggested emendations can be found in the papers in *The Chemical Revolution: Essays in Reinterpretation*, ed. Arthur Donovan, *Osiris* 4, 1988. J. B. Gough, "Lavoisier and the Fulfillment of the Stahlian Revolution," (15–33) and Robert Siegfried, "The Chemical Revolution in the History of Chemistry," (34–50) are especially relevant. A good recent summary of the historiography of emendations to the Chemical Revolution narrative is found in Frederic L. Holmes, "The Boundaries of Lavoisier's Chemical Revolution," *Revue d'histoire des sciences* 48, 1995, 11–13.

11 I.e. the phlogiston theory (although Stahl himself received virtually no historical study) and the British pneumatic chemical tradition from Stephen Hales to Priestley.

12 Herbert Butterfield, *The Origins of Modern Science* (New York: Macmillan, 1959), ch. 11. The work was first published in 1949 but the text of this chapter was changed significantly between editions.

13 Ibid., 198–99.

14 Ibid., 200. He did see the development of industrial chemistry and pharmacology as positive developments. Interestingly, the 1949 first edition lacks the entire paragraph.

[15] Thomas L. Hankins, *Science and the Enlightenment* (Cambridge: Cambridge University Press, 1985), 81. By this, he meant that chemistry was largely ancillary to medicine and pharmacy.

[16] The phrase "invention of chemistry as a discipline" is on 153. As Frederic L. Holmes noted, a work prior to Hannaway's had identified the emergence in the early eighteenth century of "a *science* of chemistry ... from the matrix of natural philosophy, medicine, alchemy, and technology." This was Robert P. Multhauf's *The Origins of Chemistry* (New York: Franklin Watts, 1966), 349. See Frederic L. Holmes, "Concepts, Operations, and the Problem of 'Modernity' in Early Modern Chemistry," *Fundamental Concepts of Early Modern Chemistry in the Context of the Operational and Experimental Practice* (Berlin: Max Planck Institute, 1995 [Preprint 25]), 50.

[17] Allen G. Debus, *The English Paracelsians* (New York: F. Watts, 1965); Betty Jo Teeter Dobbs, *The Foundation of Newton's Alchemy, or "The Hunting of the Green Lyon"* (Cambridge: Cambridge University Press, 1975), to cite two classic texts. In the 1990s and later, research on the history of seventeenth-century "chymistry" (chemistry and alchemy) was greatly augmented by the publications of Wiliam Newman and Lawrence Principe.

[18] Recently, William Newman and Lawrence Principe have argued that a *methodological* tradition anticipating many features of late eighteenth-century – and even Lavoisian – chemistry originated with Joan Baptista Van Helmont and was extended by George Starkey ("Eirenaeus Philalethes"); *Alchemy Tried in the Fire: Starkey, Boyle and the Fate of Helmontian Chymistry* (Chicago: University of Chicago Press, 2002).

[19] Arnold Thackray, *Atoms and Powers: An Essay on Newtonian Matter – Theory and the Development of Chemistry* (Cambridge, MA: Harvard University Press, 1970).

[20] By the 1990s, interest in eighteenth-century chemical application re-emerged in, for instance, Jan Golinski, *Science as Public Culture: Chemistry and Enlightenment in Britain, 1760–1820* (Cambridge: Cambridge University Press, 1992).

[21] Frederic L. Holmes, "Analysis by Fire and Solvent Extractions: The Metamorphosis of a Tradition," *Isis* 62, 1971, 129–48; *Lavoisier and the Chemistry of Life: An Exploration of Scientific Creativity* (Madison, WI: University of Wisconsin Press, 1985).

[22] Frederic Lawrence Holmes, *Eighteenth-Century Chemistry as an Investigative Enterprise* (Berkeley, CA: Office of the History of Science, 1989), 3.

[23] Ibid., 6.

[24] Ibid., 47. Holmes contrasted this with the German university pedagogical objective, e.g. of Georg Ernst Stahl.

[25] Ibid., 83.

[26] Ibid., 101. In the concluding lecture, Holmes returned to the Chemical Revolution itself to say that "If my portrayal of earlier eighteenth century chemistry is valid, then the chemical revolution cannot have overturned the science of chemistry as a whole, or transformed certain extensive areas of a science whose scope exceeded those areas." 107.

[27] Ursula Klein, "Experimental Practice and Layers of Knowledge in Early Modern Chemistry I, II and III," 73–127 in *Fundamental Concepts of Early Modern Chemistry*. She sees the emergence of this view of chemical combination as associated with changes in pharmaceutical practices of drug preparation, particularly the substitute of inorganic materials prepared by dissolution rather than the earlier dry distillation of organic substances.

[28] *Affinity, That Elusive Dream: A Genealogy of the Chemical Revolution* (Cambridge, MA: MIT Press, 2003).

[29] Berkeley: University of California Press, 1982.

[30] Edinburgh: University Press, 1975.

[31] Kim comes closest in this regard. Although incorporating the standard dramatic narrative of the Chemical Revolution into her own narrative, Kim tries to embed that narrative into a much vaster one and, at the same time, to de-center it from its dominant position.

[32] Trevor Levere's paper touches on Thomas Beddoes' conversion to anti-phlogistic chemistry.

[33] B. S. Capp, *English Almanacs, 1500–1800: Astrology and the Popular Press* (Ithaca, NY: Cornell University Press, 1979). See also Margaret C. Jacob, *Scientific Culture and the Making of the Industrial West* (New York: Oxford University Press, 1997), especially 87–96; Michael Heyd, *Be Sober and Reasonable: The Critique of Enthusiasm in the Seventeenth and Early Eighteenth Centuries* (Leiden: E.J. Brill, 1995).

[34] Principe, in his opening essay to this volume, suggests something similar to my speculation when he notes that the denigration of alchemy was particularly strong in the French and British national scientific societies "concerned about their corporate image, and thus the status of chemistry and chemists."

[35] Multhauf discusses many of these chemical crafts in *Origins of Chemistry*.

[36] Allen G. Debus, "Chemistry and the Universities in the Seventeenth Century," *Academiae Analecta*, Mededelingen van de Koninklijke Academie voor Wetenschappen Letteren en Schone Kunsten van Belgie, Klasse der Wetenschappen, 48, no. 4, 1986, 15–33; Christoph Meinel, "*Artibus Academicis Inserenda*: Chemistry's Place in Eighteenth and Early Nineteenth Century Universities," *History of Universities* 7, 1988, 89–115.

[37] "Subjection of nature's forces to man, machinery, *application of chemistry to industry and agriculture*, steam navigation, railways, electric telegraphs, clearing of whole continents for cultivation, canalization or rivers, whole populations conjured out of the ground – what earlier century had even a presentiment that such productive forces slumbered in the lap of social labor?" *Communist Manifesto*, quoted in: http://www.anu.edu.au/polsci/marx/classics/manifesto.html. My italics.

[38] *Chymia* 5, 1959, 73–112.

[39] Guerlac, *Lavoisier*, xvii.

[40] A classic example is A.E. Musson and Eric Robinson, *Science and Technology in the Industrial Revolution* (Toronto: University of Toronto Press, 1969).

[41] Karl Wilhelm Scheele is virtually the only exception.

[42] In the most recent survey of the history of eighteenth-century chemical composition, the German tradition of analytical chemistry gets not a mention. Robert Siegfried, *From Elements to Atoms: A History of Chemical Composition*, *Transactions of the American Philosophical Society* 92, pt. 4, 2002.

[43] Bernadette Bensaude-Vincent and Isabelle Stengers, *Histoire de la Chimie* (Paris: Éditions la Découverte, 1993), p. 61.

[44] A brilliant case study of this process is found in Jonathan Simon, *Chemistry, Pharmacy, and Revolution in France, 1777–1809* (Aldershot, UK: Ashgate, 2005).

INDEX

Archimedes

NEW STUDIES IN THE HISTORY AND PHILOSOPHY OF SCIENCE AND TECHNOLOGY

---

1. J.Z. Buchwald (ed.): Scientific Credibility and Technical Standards in 19th and Early *20th Century Germany and Britain*. 1996 ISBN 0-7923-4241-0
2. K. Gavroglu (ed.): *The Sciences in the European Periphery During the Enlightenment*. 1999 ISBN 0-7923-5548-2; Pb 0-7923-6562-1
3. P. Galison and A. Roland (eds.): *Atmospheric Flight in the Twentieth Century*, 2000 ISBN 0-7923-6037-0; Pb 0-7923-6742-1
4. J.M. Steele: *Observations and Predictions of Eclipse Times by Early Astronomers*. 2000 ISBN 0-7923-6298-5
5. D-W. Kim: *Leadership and Creativity.* A History of the Cavendish Laboratory, 1871–1919. 2002 ISBN 1-4020-0475-3
6. M. Feingold: *The New Science and Jesuit Science: Seventeenth Century Perspective*. 2002 ISBN 1-4020-0848-1
7. F.L. Holmes, J. Renn, H-J. Rheinberger: *Reworking the Bench*. 2003 ISBN 1-4020-1039-7
8. J. Chabás, B.R. Goldstein: *The Alfonsine Tables of Toledo*. 2003 ISBN 1-4020-1572-0
9. F.J. Dijksterhuis: *Lenses and Waves*. Christiaan Huygens and the Mathematical Science of Optics in the Seventeenth Century. 2004 ISBN 1-4020-2697-8
10. L. Corry: *David Hilbert and the Axiomatization of Physics (1898–1918)*. From Grundlagen der Geometrie to Grundlagen der Physik. 2004 ISBN 1-4020-2777-X
11. J.Z. Buchwald and A. Franklin (eds.): *Wrong for the Right Reasons*. 2005 ISBN 1-4020-3047-9
12. M. Feingold and V. Navarro-Brotons (eds.): *Universities and Science in the Early Modern Period*. 2006 ISBN 1-4020-3974-3
13. R.R. Hamerla: *An American Scientist on the Research Frontier*. Edward Morley, Community, and Radical Ideas in Nineteenth-Century Science. 2006 ISBN 1-4020-4088-1
14. J. Schickore and F. Steinle (eds.): *Revisiting Discovery and Justification*. Historical and Philosophical Perspectives on the context distinction. 2006 ISBN 1-4020-4250-7
15. K. Nickelsen: *Draughtsmen, Botanists and Nature*. The construction of Eighteenth-Century Botanical Illustrations. 2006 ISBN 1-4020-4819-X
16. R. MacLeod and J.A. Johnson (eds.): *Frontline and Factory*. Comparative Perspectives on the Chemical Industry at War, 1914–1924. 2007 ISBN 1-4020-5489-0
17. G. Schiemann (ed.): *Herman von Helmholtz's Mechanism at the Dawn of Modernity*. A Study on the Transition from Classical to Modern Philosophy of Nature. 2008 ISBN 1-4020-5629-1
18. L.M. Principe (ed.): *New Narratives in Eighteenth-Century Chemistry*. Contributions from the First Francis Bacon Workshop, 21–23 April 2005. 2007 ISBN 978-1-4020-6273-5

---

springer.com

Printed in the United States
90336LV00003BA/3/A

9 781402 062735